Heat Recovery Steam Generators

Thermal Design & Testing

by D. James Benton

Foreword

Heat recovery steam generators (HRSGs) are an integral part of any modern combined cycle power plant. These heat exchangers are designed to recover the heat from a gas turbine exhaust and convert this to steam, which drives a turbine and ultimately a second generator. Because the gas turbine operates efficiently at high temperatures and heat is removed down to nearly ambient temperature, the overall efficiency of this complex design exceeds that of conventional Rankine cycle coal-, oil-, or gas-fired systems. These systems can also be operated in simple or combined cycle, increasing flexibility and response to load dispatch. HRSG can be complex to analyze and also difficult to effectively test and prove their performance. This text covers both theory and testing in practice with examples.

All of the examples contained in this book,
(as well as a lot of free programs) are available at...

https://www.dudleybenton.altervista.org/software/index.html

Example spreadsheets are provided
in both SI and English units.

Trademarks

Excel® is a registered trademark of Microsoft®, as is the company name. GateCycle™ is a trademark of the General Electric Company® and the name of a very useful thermal cycle modeling tool.

i

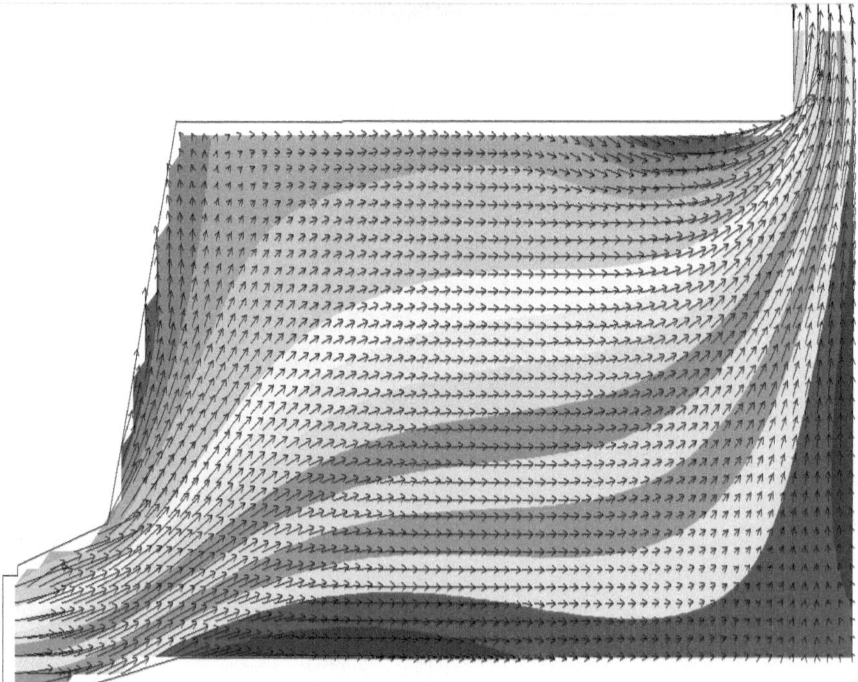

Table of Contents   page

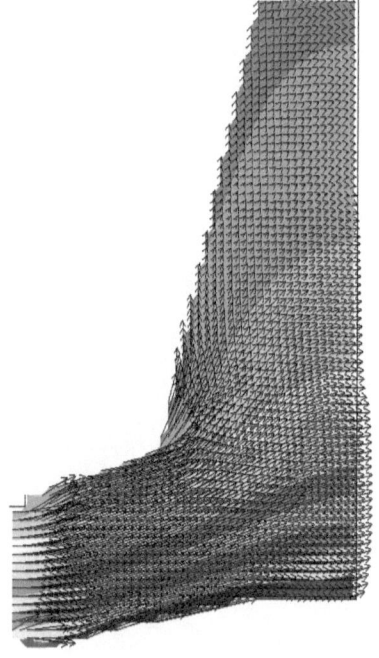

Chapter 1. Introduction

In this text we will consider various aspects of design and testing heat recovery steam generators (HRSGs). Steam properties are covered in Appendix A and exhaust gas properties in Appendix B. Combustion, in so far as it applies to HRSGs, is covered in Appendix C. Design is divided into two sections: 1) individual components and 2) splitting and placement along the gas stream. Performance calculations and testing are covered after the design sections.

Orientation

The elements (components) of a HRSG (i.e., economizers, evaporators, and superheaters) are crossflow tube-in-shell heat exchangers, having steam (liquid or vapor) on the inside and exhaust gas on the outside. While some components may be designated as *counterflow* by some manufacturers (e.g., Nooter-Eriksen), this is merely conceptual or for the purposes of calculation. Nobody puts horizontal tube banks in a HRSG running counter-current to the gas flow. You might choose to perform heat transfer calculations as if such a component were counterflow, but be aware that this is not physically the case with any.

Phase Change

HRSGs are somewhat unusual in that they are designed specifically to separate the phase-change and non-phase-change sections. It is assumed (and forced in GateCycle™) that there be no evaporation in an economizer or superheater. That is, liquid enters and leaves any and all economizers, while dry (saturated or superheated) vapor enters and leaves any and all superheaters. All of the phase change occurs in the evaporators. GateCycle™ allows subcooled liquid at the inlet of an evaporator, but only saturated vapor at the exit. This means that, except at the inlet of the evaporator, there is no phase-change anywhere except in the evaporators and there is only phase change in the evaporators (apart from the initial heating, which may or may not be in physically separate elements). All of the manufacturers that I have worked with follow this same pattern. Even what are called *once-through* boilers typically have three sections (economizer, evaporator, and superheater). Because of this peculiarity, heat transfer calculations within HRSG components are more specific than for heat exchangers in general.

1

Chapter 2. Analytical vs. Numerical

As mentioned in Chapter 1, HRSG elements are crossflow tube-in-shell heat exchangers; so we will first consider the available methods for calculating heat transfer in such devices. There are three primary methods: 1) NTU-effectiveness, 2) F-LMTD, and 3) numerical. All three methods can be found in most heat transfer texts, for example the one by Lindon C. Thomas.[1] The most common and easily implemented numerical method is finite difference, which can be found in Appendix P of Thomas' professional version textbook. We will compare these three methods for several cases.

The NTU-Effectiveness Method

This method is similar to the LMTD method, but takes a slightly different approach to solving the same differential equation. The effectiveness, P, is defined as the ratio of the actual to the maximum heat transfer:

$$P = \frac{Q}{Q_{MAX}} \tag{2.1}$$

The maximum possible heat transfer would be equal to the difference in the two inlet temperatures times the minimum product of the mass flow and specific heat or:

$$Q_{MAX} = (\dot{m}C_P)_{MIN}(T_{H,in} - T_{C,in}) \tag{2.2}$$

where the minimum could be the hot or cold side. The ratio of the minimum to maximum product of mass and specific heat is given the symbol, R:

$$R = \frac{(\dot{m}C_P)_{MIN}}{(\dot{m}C_P)_{MAX}} \tag{2.3}$$

The number of transfer units, NTU, is defined as:

$$NTU = \frac{UA}{(\dot{m}C_P)_{MIN}} \tag{2.4}$$

The formula relating P(NTU,R) may be found in Appendix N-1 of the previous reference and in the spreadsheet crossflow1.xls.

$$P = \frac{1 - e^{-\Gamma R}}{R} \tag{2.5}$$

$$\Gamma = 1 - e^{-NTU} \tag{2.6}$$

---

[1] https://www.amazon.com/Lindon-C.-Thomas/e/B001HOVPI4

The F-LMTD Method

The heat transfer (Q) is the product of the overall conductance (UA), the log-mean temperature difference (LMTD), and the fudge factor (F):

$$Q = UA \times F \times LMTD \tag{2.7}$$

The factor F for one side mixed (outside the tubes) and one side unmixed (inside the tubes) in any heat transfer text. Most often this is given graphically, as in the following figure:

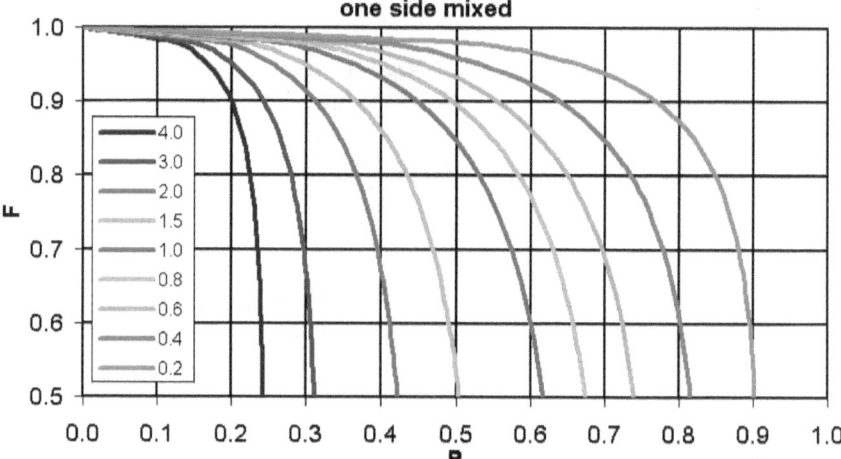

This figure appears in Section 11-8-1 of Thomas' text and was originally published by Bowman, Mueller, and Nagle.[2] The figure along with the formulas can be found in the online archive in folder examples\crossflow in spreadsheet crossflow1.xls. The factor F is given by the following relationship with Γ as before from Equation 2.6:

$$F = \frac{\ln\left(\dfrac{R\,e^{-\Gamma R}}{R - 1 + e^{-\Gamma R}}\right)}{NTU(1 - R)} \tag{2.8}$$

If R=0, this reduces to:

$$F = \frac{1 - e^{-\Gamma}}{NTU\, e^{-\Gamma}} \tag{2.9}$$

---

[2] Bowman, R. A., Mueller, A. C., and Nagel, W. M., "Mean Temperature Difference in Heat Exchanger Design," Transactions of the ASME, No. 62, p. 283, 1940.

The Finite Difference Numerical Method

The first step of implementing the Finite Difference Method (FDM) is to break the domain up into smaller cells (or elements) and to consider the heat transfer within one of these cells. The physical orientation is immaterial.

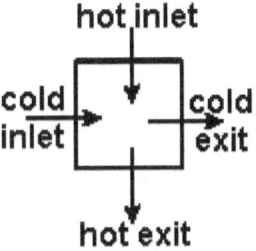

The fully explicit method (often called Euler's) considers only the inlet temperatures; thus the heat transfer within this cell is given by:

$$\Delta Q = \Delta UA\left(T_{HOT,IN} - T_{COLD,IN}\right) \qquad (2.10)$$

If the domain is broken up into a uniform rectangular grid of nx*ny cells and the overall heat transfer coefficient, U, is presumed to be uniform over the heat transfer surface, $\Delta UA = UA/nx/ny$. The conservation of energy (1st Law of Thermodynamics) for this cell is:

$$\Delta Q = \Delta m_{HOT} C_{P,HOT}\left(T_{HOT,IN} - T_{HOT,OUT}\right) \qquad (2.11)$$

$$\Delta Q = \Delta m_{COLD} C_{P,COLD}\left(T_{COLD,OUT} - T_{COLD,IN}\right) \qquad (2.12)$$

For now the specific heats, C_P, will be considered constant. We will next consider variable properties. The cell mass flows are presumed equally distributed: $\Delta m_{HOT} = m_{HOT}/nx$, $\Delta m_{COLD} = m_{COLD}/ny$.

Example 2.1

All three calculations are implemented in the spreadsheet crossflow1.xls, which is based on Example 11-21 in Thomas' text. Hot water enters from the top at 65°C and a rate of 5.25 kg/s. Cold air enters from the left side at 15°C and a rate of 2.39 kg/s. The specific heats are 4.19 and 1.01 kJ/kg/°C for the hot and cold sides, respectively. The overall heat transfer coefficient, U, is 205 W/m²/°C and the total surface area is 5.98 m². The P-NTU method can be calculated based on either the hot or cold side and so it is implemented both ways to illustrate that the same end result is obtained. The F-LMTD has only one variant, which produces the same result. The calculations are exactly as indicated by Equations 2.1 through 2.9. Inputs and outputs are shown in the following figure, which is taken directly from the spreadsheet:

	A	B	C	D
1	**crossflow example 11-211**			
2	symbol	units	hot	cold
3	U	W/m²/°C	205	
4	Cp	kJ/kg/°C	**4.19**	**1.01**
5	m	kg/s	**5.25**	**2.39**
6	Tin	°C	**65.0**	**15.0**
7	Tout	°C	**62.9**	34.5
8	q	kW	47.1	
9	LMTD	°C	38.52	
10	P	-	0.043	0.390
11	R	-	9.113	0.110
12	symbol	units	one mixed	
13	F-LMTD method			
14	F	-	0.995	
15	UA	kW/°C	1.23	
16	A	m²	5.98	
17	P-NTU method			
18	NTU	-	0.0558	0.509
19	UA	kW/°C	1.23	1.23
20	A	m²	5.98	5.98
21	**user inputs in blue**			
22	calculations in orange			
23	**linked cells in green**			

The hot and cold temperatures are shown in this next figure:

Q	R	S	T	U	V	W	X	Y	Z	AA	AB	AC
	crossflow (hot mixed/cold unmixed) - explicit - 10x10 grid											
1.23						hot side inlet						
		65.0	65.0	65.0	65.0	65.0	65.0	65.0	65.0	65.0	65.0	
		64.8	64.8	64.8	64.8	64.8	64.8	64.8	64.8	64.8	64.8	
		64.6	64.6	64.6	64.6	64.6	64.6	64.6	64.6	64.6	64.6	
		64.3	64.3	64.3	64.3	64.3	64.3	64.3	64.3	64.3	64.3	
		64.1	64.1	64.1	64.1	64.1	64.1	64.1	64.1	64.1	64.1	
		63.9	63.9	63.9	63.9	63.9	63.9	63.9	63.9	63.9	63.9	
		63.7	63.7	63.7	63.7	63.7	63.7	63.7	63.7	63.7	63.7	
		63.5	63.5	63.5	63.5	63.5	63.5	63.5	63.5	63.5	63.5	
		63.2	63.2	63.2	63.2	63.2	63.2	63.2	63.2	63.2	63.2	
	Texit	63.0	63.0	63.0	63.0	63.0	63.0	63.0	63.0	63.0	63.0	
	62.8	62.8	62.8	62.8	62.8	62.8	62.8	62.8	62.8	62.8	62.8	
						hot side outlet						
	15.0	17.5	20.0	22.2	24.4	26.5	28.4	30.3	32.1	33.7	35.3	
	15.0	17.5	19.9	22.2	24.4	26.4	28.4	30.2	32.0	33.7	35.2	
	15.0	17.5	19.9	22.2	24.3	26.4	28.3	30.2	31.9	33.6	35.2	
cold side inlet	15.0	17.5	19.9	22.2	24.3	26.3	28.3	30.1	31.8	33.5	35.1	cold side outlet
	15.0	17.5	19.9	22.1	24.3	26.3	28.2	30.0	31.8	33.4	35.0	
	15.0	17.5	19.8	22.1	24.2	26.2	28.1	30.0	31.7	33.3	34.9	
	15.0	17.5	19.8	22.1	24.2	26.2	28.1	29.9	31.6	33.2	34.8	
	15.0	17.5	19.8	22.0	24.1	26.1	28.0	29.8	31.5	33.2	34.7	
	15.0	17.5	19.8	22.0	24.1	26.1	28.0	29.8	31.5	33.1	34.6	
	15.0	17.4	19.8	22.0	24.1	26.0	27.9	29.7	31.4	33.0	34.5	
											Texit	**34.9**

6

Notice that the hot temperatures are mixed (i.e., the same horizontally) and the cold temperatures are not mixed (i.e., vary horizontally and vertically) The temperature differences (fully explicit) and heat transfer for each cell are shown in this next figure:

AD	AE	AF	AG	AH	AI	AJ	AK	AL	AM
crossflow - explicit - 10x10 grid									
ΔT in each cell									
use max(0,dT) to prevent overshoot									
50.0	47.5	45.0	42.8	40.6	38.5	36.6	34.7	32.9	31.3
49.8	47.2	44.8	42.6	40.4	38.3	36.4	34.5	32.8	31.1
49.6	47.0	44.6	42.4	40.2	38.2	36.2	34.4	32.6	31.0
49.3	46.8	44.4	42.2	40.0	38.0	36.1	34.2	32.5	30.8
49.1	46.6	44.2	42.0	39.9	37.8	35.9	34.1	32.3	30.7
48.9	46.4	44.0	41.8	39.7	37.7	35.7	33.9	32.2	30.6
48.7	46.2	43.9	41.6	39.5	37.5	35.6	33.8	32.1	30.4
48.5	46.0	43.7	41.4	39.3	37.3	35.4	33.6	31.9	30.3
48.2	45.8	43.5	41.2	39.2	37.2	35.3	33.5	31.8	30.2
48.0	45.6	43.3	41.1	39.0	37.0	35.1	33.3	31.6	30.0
ΔQ in each cell									
0.61	0.58	0.55	0.52	0.50	0.47	0.45	0.43	0.40	0.38
0.61	0.58	0.55	0.52	0.50	0.47	0.45	0.42	0.40	0.38
0.61	0.58	0.55	0.52	0.49	0.47	0.44	0.42	0.40	0.38
0.61	0.57	0.55	0.52	0.49	0.47	0.44	0.42	0.40	0.38
0.60	0.57	0.54	0.52	0.49	0.46	0.44	0.42	0.40	0.38
0.60	0.57	0.54	0.51	0.49	0.46	0.44	0.42	0.40	0.38
0.60	0.57	0.54	0.51	0.49	0.46	0.44	0.41	0.39	0.37
0.59	0.56	0.54	0.51	0.48	0.46	0.43	0.41	0.39	0.37
0.59	0.56	0.53	0.51	0.48	0.46	0.43	0.41	0.39	0.37
0.59	0.56	0.53	0.50	0.48	0.45	0.43	0.41	0.39	0.37
								q	48.1

The calculated exit hot temperature (bold number below Texit on the left side of the second figure, half-way down) is 62.8°C, close, but not exactly equal, to the analytical result, 62.9°C (cell C7 in the first figure). The calculated exit cold temperature (bold number beside Texit at the bottom right of the second figure) is 34.9°C, close, but not exactly equal, to the analytical result, 34.9°C (cell D7 in the first figure). The calculated total heat transfer (bold number beside q in the bottom right corner of the third figure) is 48.1 kW is close, but not exactly equal, to the analytical result, 47.1 kW (cell C8 in the first figure).

Discretization

While we could use more cells, reducing the size of each one and increasing the discretization of the FDM calculations in the Excel® spreadsheet, this would be both inefficient and unnecessary. Instead, we will implement the entire calculation in C. The source code (crossflow1.x) as well as a little batch file to compile it (_compile.bat) can be found in this same folder. The analytical functions are very similar to the VBA code in the spreadsheet:

```
double F1mix(double R,double NTU)
   {
   double G;
   G=1.-exp(-NTU);
   if(fabs(R-1.)<0.01)
     return((1.-exp(-G))/exp(-G)/NTU);
   return(log(R*exp(-G*R)/(R-1.+exp(-G*R)))/NTU/(1.-R));
   }

double NTUofF(double R,double F)
   {
   int iter;
   double N1,N2,NTU;
   N1=0.01;
   N2=100.;
   for(iter=0;iter<32;iter++)
     {
     NTU=(N1+N2)/2.;
     if(F1mix(R,NTU)>F)
       N1=NTU;
     else
       N2=NTU;
     }
   return(NTU);
   }

double P1mix(double R,double NTU)
   {
   double G;
   G=1.-exp(-NTU);
   return((1.-exp(-G*R))/R);
   }

double NTU1mix(double R,double P)
   {
   int iter;
   double N1,N2,NTU;
   N1=0.01;
   N2=100.;
   for(iter=0;iter<32;iter++)
     {
     NTU=(N1+N2)/2.;
     if(P1mix(R,NTU)<P)
       N1=NTU;
     else
       N2=NTU;
     }
   return(NTU);
   }
```

```
double fLMTD(double dT1,double dT2)
    {
    if(dT1<=0.||dT2<=0.)
      return(0.);
    if(fabs(dT1-dT2)<0.01)
      return(sqrt(dT1*dT2));
    return((dT1-dT2)/log(dT1/dT2));
    }
```

The analytical results are duplicated, as shown below:

```
crossflow example 1
62.86 hot side outlet temperature [°C]
34.51 cold side outlet temperature [°C]
0.0428 hot side P []
9.1128 hot side R []
0.0558 hot side NTU []
1.228 hot side UA [kW/°C]
5.989 hot side A [m²]
0.3902 cold side P []
0.1097 cold side R []
0.5089 cold side NTU []
1.228 cold side UA [kW/°C]
5.992 cold side A [m²]
38.15 LMTD [°C]
0.9953 F []
1.240 UA [kW/°C]
6.050 A [m²]
```

The FDM results are implemented in a function that accepts grid size as an input so that we can investigate the convergence:

```
void FDM(int nx,int ny,int list)
    {
    int i,j;
    double dQ,dT,Qh,*Tc,*Th;
    Tc=calloc(ny*(nx+1),sizeof(double));
    Th=calloc(ny+1,sizeof(double));
    Th[0]=Thi;
    for(i=0;i<ny;i++)
      Tc[(nx+1)*i]=Tci;
    for(Tco=i=0;i<ny;i++)
      {
      for(Qh=j=0;j<nx;j++)
        {
        dT=fmax(0.,Th[i]-Tc[(nx+1)*i+j]);
        dQ=UA*dT/nx/ny;
        Qh+=dQ;
        Tc[(nx+1)*i+j+1]=Tc[(nx+1)*i+j]+dQ/(mC/ny)/CpC;
        }
      Tco+=Tc[(nx+1)*i+nx];
```

```
      Th[i+1]=Th[i]-Qh/mH/CpH;
        }
    Tco/=ny;
    Tho=Th[ny];
    Q=(mH*CpH*(Thi-Tho)+mC*CpC*(Tco-Tci))/2.;
      }
```

Notice that it's only necessary to allocate Th[ny+1] and Tc[ny*(nx+1)]. The extra row (ny+1) or column (nx+1) in these arrays provides the cold and hot inlet conditions, respectively. The same number of cells is present in the spreadsheet. Because the hot side is mixed, it is not necessary to allocate a two-dimensional array, only a one-dimensional. The output for 10x10 cells is:

```
<------------------------Tcold-------------------->Thot
17.6,20.0,22.3,24.5,26.6,28.6,30.4,32.2,33.9,35.5,64.8
17.6,20.0,22.3,24.5,26.5,28.5,30.4,32.1,33.8,35.4,64.6
17.5,20.0,22.3,24.4,26.5,28.4,30.3,32.1,33.7,35.3,64.3
17.5,19.9,22.2,24.4,26.4,28.4,30.2,32.0,33.6,35.2,64.1
17.5,19.9,22.2,24.3,26.4,28.3,30.2,31.9,33.6,35.1,63.9
17.5,19.9,22.2,24.3,26.3,28.3,30.1,31.8,33.5,35.0,63.7
17.5,19.9,22.1,24.3,26.3,28.2,30.0,31.8,33.4,34.9,63.4
17.5,19.9,22.1,24.2,26.2,28.1,30.0,31.7,33.3,34.9,63.2
17.5,19.8,22.1,24.2,26.2,28.1,29.9,31.6,33.2,34.8,63.0
17.5,19.8,22.0,24.1,26.1,28.0,29.8,31.5,33.1,34.7,62.8
nx=10, ny=10, Q=48.48, Tho=62.80, Tco=35.09
```

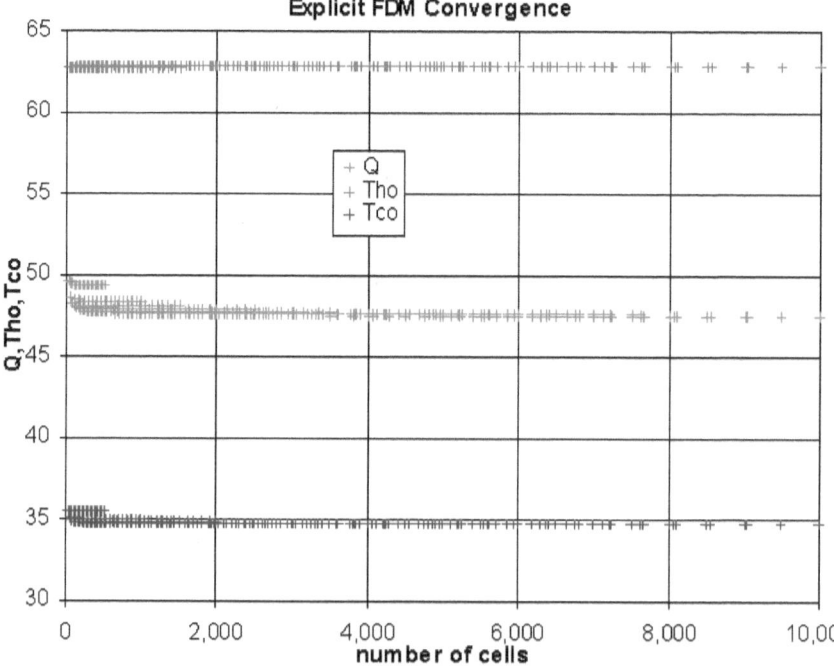

Explicit FDM Convergence

10

The explicit FDM convergence (to Q, Tho, and Tco) is illustrated in the preceding figure, which is also included in the spreadsheet along with the output of the program. It takes at least 2,000 cells ($\approx$45x45) to converge (i.e., to obtain values that are independent of the number of cells). This may seem like an extreme result, but is not uncommon for explicit methods, which is why implicit methods are often employed. Implicit methods use the exiting as well as entering temperatures. As these are not known from the outset, the process is implicit and requires iterative calculation. The only alternative to iterative calculations arises when Equations 2.11, 2.12, and 2.13 can be solved simultaneously.

$$\Delta T = \frac{\left(T_{HOT,IN} - T_{COLD,IN}\right) + \left(T_{HOT,OUT} - T_{COLD,OUT}\right)}{2} \qquad (2.13)$$

This cannot be entirely implemented in this case, as the hot side is presumed to be mixed, but can be partially implemented. The approximate solution is:

$$Q = \frac{2m_C C_{PC} m_H C_{PH} (T_{HI} - T_{CI}) UA}{\left(m_C C_{PC} + m_H C_{PH}\right) UA + 2m_C C_{PC} m_H C_{PH}} \qquad (2.14)$$

The implicit variant is also implemented in the same code (crossflow1.c). The output for 10x10 is listed below:

```
implicit FDM
<---------------------Tcold--------------------->Thot
17.5,19.9,22.1,24.3,26.3,28.2,30.1,31.8,33.5,35.0,64.8
17.5,19.8,22.1,24.2,26.3,28.2,30.0,31.7,33.4,35.0,64.6
17.5,19.8,22.1,24.2,26.2,28.1,29.9,31.7,33.3,34.9,64.3
17.5,19.8,22.0,24.1,26.2,28.1,29.9,31.6,33.2,34.8,64.1
17.5,19.8,22.0,24.1,26.1,28.0,29.8,31.5,33.2,34.7,63.9
17.4,19.8,22.0,24.1,26.1,27.9,29.7,31.5,33.1,34.6,63.7
17.4,19.7,21.9,24.0,26.0,27.9,29.7,31.4,33.0,34.5,63.5
17.4,19.7,21.9,24.0,26.0,27.8,29.6,31.3,32.9,34.4,63.3
17.4,19.7,21.9,23.9,25.9,27.8,29.5,31.2,32.8,34.4,63.1
17.4,19.7,21.8,23.9,25.9,27.7,29.5,31.2,32.8,34.3,62.8
nx=10, ny=10, Q=47.45, Tho=62.84, Tco=34.66
```

The final values for 100x100 cells are:

```
nx=100, ny=100, Q=47.44, Tho=62.84, Tco=34.65
```

The following figure shows why implicit methods are so often used:

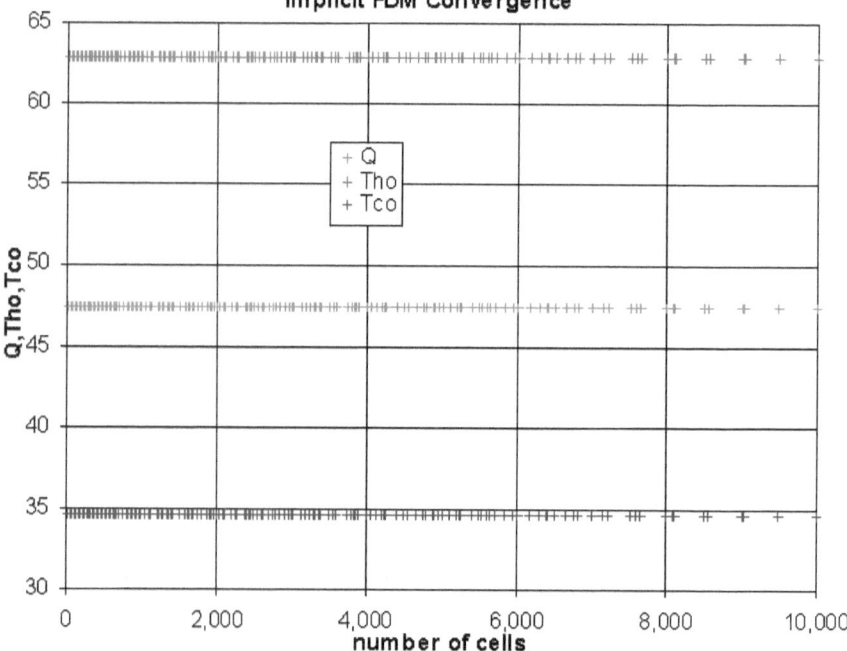

The implicit FDM converges even with 5x5 cells, an order of magnitude fewer cells and a similar reduction in calculation time. Notice that the final converged values for the FDM method (explicit or implicit) do not exactly match the analytical values for either the P-NTU or F-LMTD methods. The final values of Q are 47.05 and 47.44 kW for the analytical and numerical methods, respectively. The final values of Tho are 62.86 and 62.84°C for the analytical and numerical methods, respectively. The final values of Tco are 34.49 and 34.65°C for the analytical and numerical methods, respectively. Some of these differences are due to round off in the input parameters and the example in the textbook.

Variable Properties and Heat Transfer Coefficient

The FDM is naturally suited to variable properties and heat transfer coefficients; whereas, the analytical methods are not. If these change throughout the heat exchanger, one might achieve greater accuracy by using average values, in general, this only works when the quantities vary linearly *and* simultaneously. If C_P rises while U falls or vice versa, this is not the same as if they both rise or both fall together. Averages do not account for variation in quantities that combine multiplicatively or by division, let alone the sort of relationship in Equation 2.14. We will first consider the case of C_{PH} varying positively with temperature from 0.5 to 1.5 times the mean, such that the average is 1.0. We will also generate contours of the cold side temperatures in order to facilitate comparisons. The code (crossflow2.c) in the same folder is setup to do this conveniently. The cold side temperature contours are:

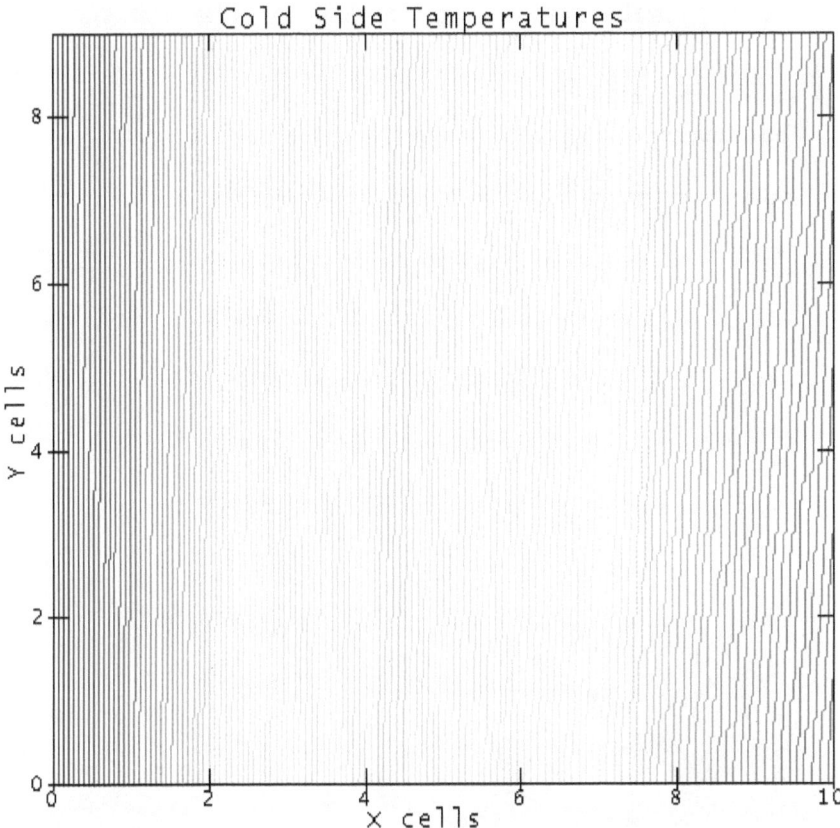

Having variable specific heat will distort the temperature profile and also change the overall result. The specific heat is:

```
double CpH(double T)/*hot side specific heat [kJ/kg°C]*/
    {
    return(4.19*(0.5+(T-62.84)/(65.-62.84)));
    }
```

The results are:

```
implicit FDM
<----------------------Tcold---------------------->Thot
17.5,19.9,22.1,24.3,26.3,28.2,30.1,31.8,33.5,35.1,64.9
17.5,19.9,22.1,24.2,26.3,28.2,30.0,31.8,33.4,35.0,64.7
17.5,19.8,22.1,24.2,26.2,28.2,30.0,31.7,33.4,34.9,64.5
```

```
17.5,19.8,22.1,24.2,26.2,28.1,29.9,31.7,33.3,34.9,64.4
17.5,19.8,22.0,24.2,26.2,28.1,29.9,31.6,33.2,34.8,64.2
17.5,19.8,22.0,24.1,26.1,28.0,29.8,31.5,33.2,34.7,64.0
17.4,19.8,22.0,24.1,26.1,28.0,29.8,31.5,33.1,34.6,63.8
17.4,19.8,22.0,24.0,26.0,27.9,29.7,31.4,33.0,34.6,63.6
17.4,19.7,21.9,24.0,26.0,27.8,29.6,31.3,32.9,34.5,63.3
17.4,19.7,21.9,23.9,25.9,27.8,29.5,31.2,32.8,34.4,63.0
variable: nx=10, ny=10, Q=46.59, Tho=63.01, Tco=34.74
constant: nx=10, ny=10, Q=47.45, Tho=62.84, Tco=34.66
```

The difference contours are shown below:

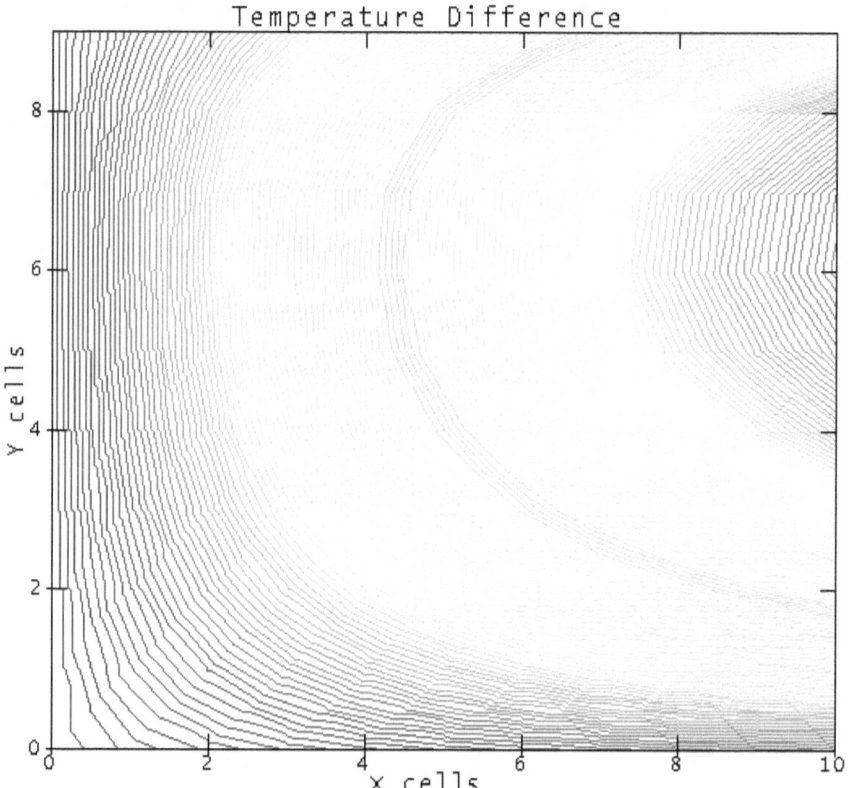

While the differences are very small (<0.123°C), they are not uniformly distributed. This example illustrates two things: 1) even changing Cp from 0.5 to 1.5 about the mean doesn't produce extreme differences and 2) the differences vary over the heat exchanger. We will now do the same thing with the cold side specific heat by modifying the same code. The specific heat function is very similar:

```
double CpC(double T)/*cold side specific ht [kJ/kg/°C]*/
   {
```

14

```
    return(1.01*(0.5+(T-15.)/(34.66-15.)));
    }
```

The output is as follows:

```
implicit FDM
<--------------------Tcold-------------------->Thot
19.9,22.9,25.2,27.1,28.8,30.4,31.7,32.9,34.1,35.1,64.8
19.9,22.8,25.2,27.1,28.8,30.3,31.7,32.9,34.0,35.1,64.6
19.8,22.8,25.1,27.1,28.8,30.3,31.6,32.8,34.0,35.0,64.4
19.8,22.8,25.1,27.0,28.7,30.2,31.6,32.8,33.9,35.0,64.2
19.8,22.7,25.1,27.0,28.7,30.2,31.5,32.7,33.9,34.9,64.0
19.8,22.7,25.0,26.9,28.6,30.1,31.4,32.7,33.8,34.8,63.8
19.8,22.7,25.0,26.9,28.6,30.1,31.4,32.6,33.7,34.8,63.6
19.7,22.7,25.0,26.9,28.5,30.0,31.3,32.6,33.7,34.7,63.4
19.7,22.6,24.9,26.8,28.5,30.0,31.3,32.5,33.6,34.7,63.2
19.7,22.6,24.9,26.8,28.4,29.9,31.2,32.5,33.6,34.6,63.0
variable: nx=10, ny=10, Q=46.65, Tho=62.95, Tco=34.87
constant: nx=10, ny=10, Q=47.45, Tho=62.84, Tco=34.66
```

The difference contours are shown below:

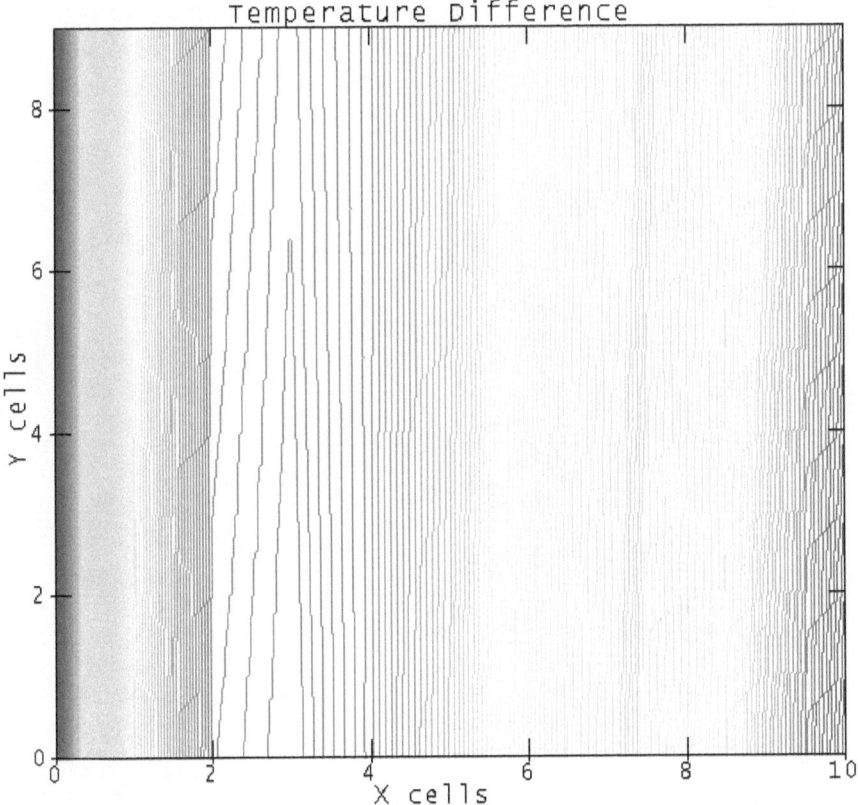

The differences resulting from variable cold side specific heat are 25 times as large as with varying the hot side specific heat by the same ratio (≤3.071°C compared to ≤0.123°C). An entirely different pattern arises also. If nothing else, this demonstrates that simply using average properties and hoping for the best is not a wise path, especially if you might face liquidated damages for inadequate performance. The temperatures themselves (rather than the differences) are very similar to the previous Cold Side Temperatures graph. We next vary both specific heats in the same way as before. The temperature differences are:

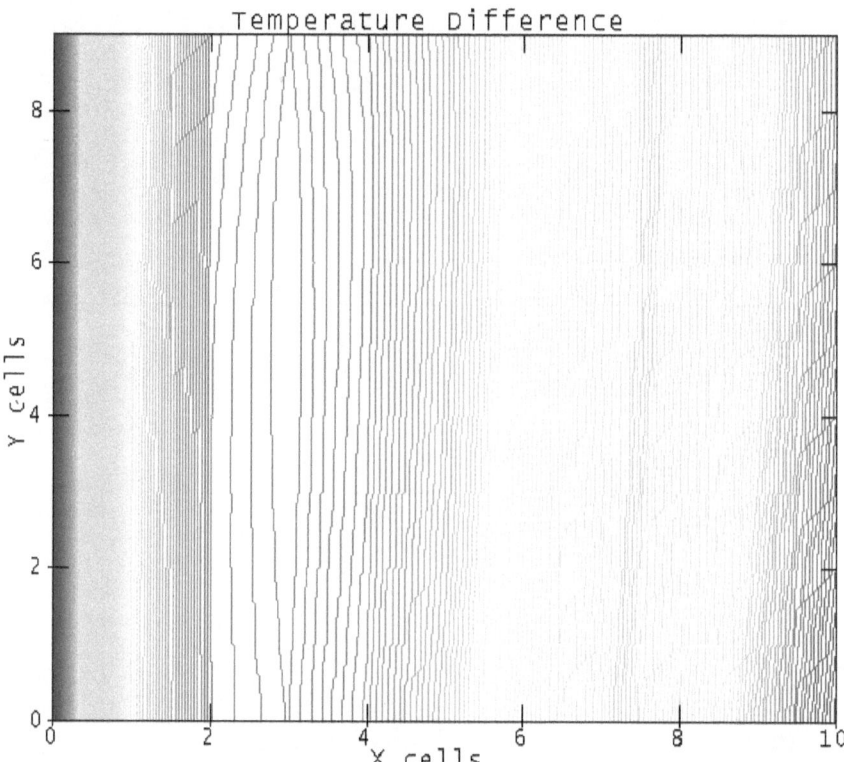

Temperature Difference

The output is:

```
implicit FDM
<----------------------Tcold---------------------->Thot
19.9,22.9,25.2,27.2,28.9,30.4,31.7,33.0,34.1,35.2,64.9
19.9,22.8,25.2,27.1,28.8,30.3,31.7,32.9,34.1,35.1,64.7
19.8,22.8,25.2,27.1,28.8,30.3,31.6,32.9,34.0,35.1,64.6
19.8,22.8,25.1,27.1,28.8,30.3,31.6,32.8,34.0,35.0,64.4
19.8,22.8,25.1,27.0,28.7,30.2,31.6,32.8,33.9,35.0,64.2
19.8,22.8,25.1,27.0,28.7,30.2,31.5,32.7,33.9,34.9,64.1
19.8,22.7,25.0,27.0,28.6,30.1,31.5,32.7,33.8,34.9,63.9
19.8,22.7,25.0,26.9,28.6,30.1,31.4,32.6,33.8,34.8,63.7
```

```
19.7,22.7,25.0,26.9,28.6,30.0,31.4,32.6,33.7,34.7,63.4
19.7,22.6,24.9,26.8,28.5,30.0,31.3,32.5,33.6,34.7,63.2
variable: nx=10, ny=10, Q=45.98, Tho=63.16, Tco=34.94
constant: nx=10, ny=10, Q=47.45, Tho=62.84, Tco=34.66
```

The maximum difference is ≤3.099°C. The pattern is similar. We note that that $m_C C_{PC}$ is much less than $m_H C_{PH}$, so that the cold side specific heat and mass flow dominate this problem. You can modify this program (crossflow2.c) to demonstrate a reflexive pattern when the hot side specific heat and mass flow dominate. We will now vary the conductance by a factor of 0.5 to 1.5 over the hot side, using constant cold and hot side Cp. The temperature difference contours are (<0.2206°C):

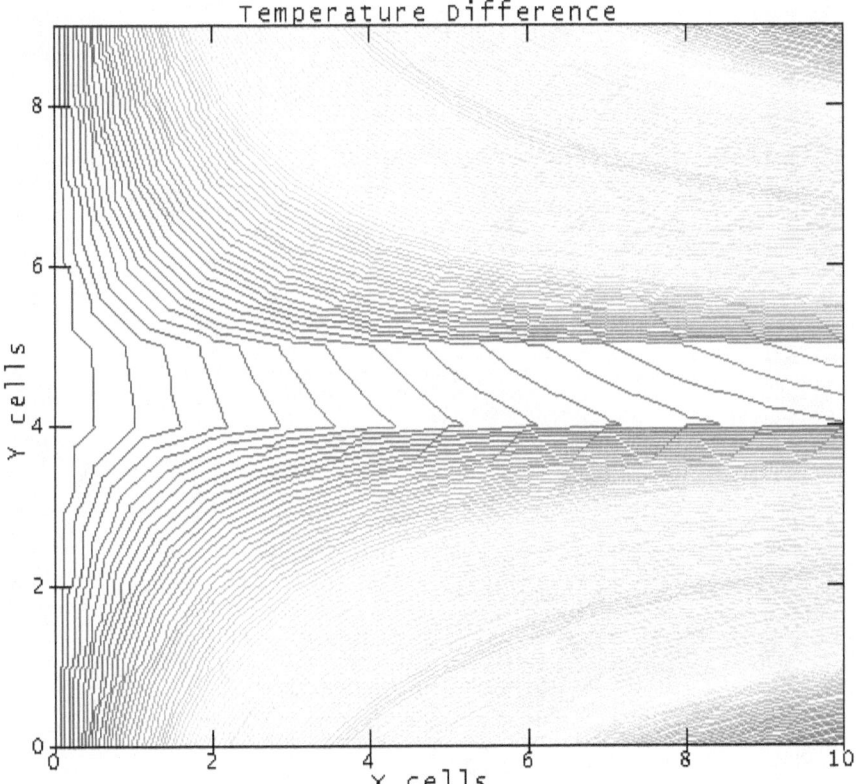

The results are:

```
implicit FDM
<----------------------Tcold---------------------->Thot
17.5,19.9,22.2,24.4,26.4,28.4,30.3,32.0,33.7,35.3,64.8
17.5,19.9,22.2,24.3,26.4,28.3,30.1,31.9,33.6,35.1,64.6
17.5,19.9,22.1,24.3,26.3,28.2,30.0,31.8,33.4,35.0,64.3
17.5,19.8,22.1,24.2,26.2,28.1,29.9,31.7,33.3,34.9,64.1
```

```
17.5,19.8,22.0,24.1,26.1,28.0,29.8,31.5,33.2,34.7,63.9
17.4,19.8,22.0,24.1,26.0,27.9,29.7,31.4,33.0,34.6,63.7
17.4,19.7,21.9,24.0,26.0,27.8,29.6,31.3,32.9,34.5,63.5
17.4,19.7,21.9,23.9,25.9,27.7,29.5,31.2,32.8,34.3,63.3
17.4,19.7,21.8,23.9,25.8,27.7,29.4,31.1,32.7,34.2,63.1
17.4,19.6,21.8,23.8,25.7,27.6,29.3,31.0,32.6,34.1,62.8
variable: nx=10, ny=10, Q=47.45, Tho=62.84, Tco=34.66
constant: nx=10, ny=10, Q=47.45, Tho=62.84, Tco=34.66
```

The overall differences are within 0.01°C and the maximum difference is 0.2206°C. Varying UA over the cold side yields the following (0.08°C):

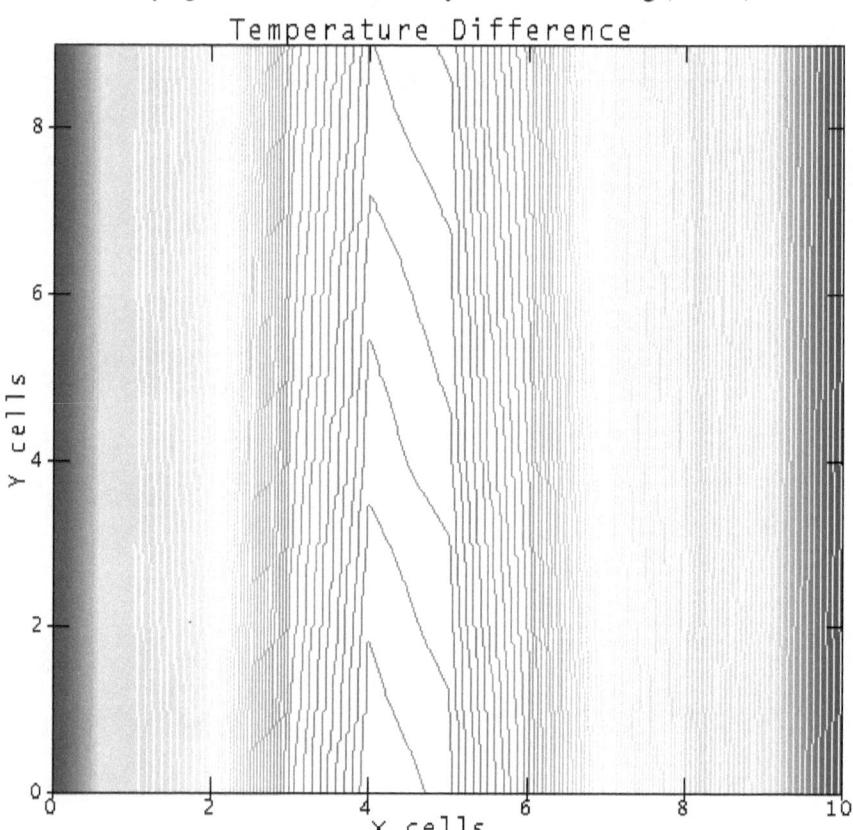

The results are:

```
implicit FDM
<----------------------Tcold---------------------->Thot
17.5,19.9,22.2,24.3,26.4,28.3,30.1,31.9,33.5,35.1,64.8
17.5,19.9,22.2,24.3,26.3,28.2,30.1,31.8,33.4,35.0,64.6
17.5,19.9,22.1,24.3,26.3,28.2,30.0,31.7,33.3,34.9,64.3
17.5,19.9,22.1,24.2,26.2,28.1,29.9,31.6,33.3,34.8,64.1
```

```
17.5,19.8,22.1,24.2,26.2,28.1,29.9,31.6,33.2,34.7,63.9
17.5,19.8,22.0,24.1,26.1,28.0,29.8,31.5,33.1,34.6,63.7
17.5,19.8,22.0,24.1,26.1,28.0,29.7,31.4,33.0,34.5,63.5
17.5,19.8,22.0,24.1,26.0,27.9,29.7,31.3,32.9,34.4,63.3
17.4,19.8,21.9,24.0,26.0,27.8,29.6,31.3,32.9,34.4,63.1
17.4,19.7,21.9,24.0,25.9,27.8,29.5,31.2,32.8,34.3,62.8
variable: nx=10, ny=10, Q=47.45, Tho=62.84, Tco=34.66
constant: nx=10, ny=10, Q=47.45, Tho=62.84, Tco=34.66
```

Varying both specific heats and UA over the cold (steam) side, which is more likely, yields (<3.2275°C):

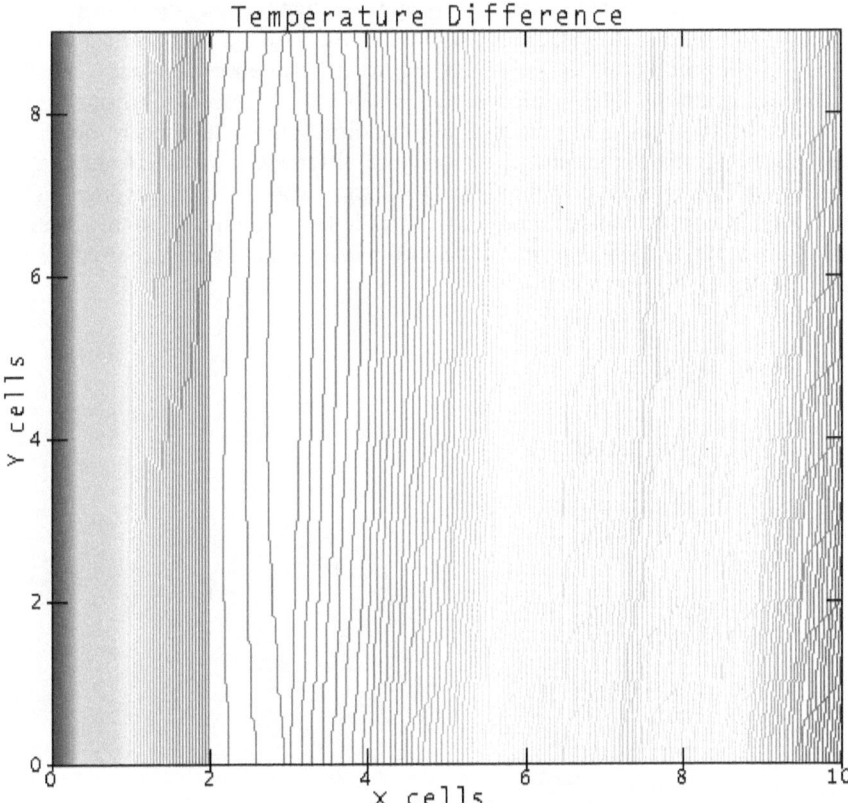

The results are:

```
implicit FDM
<------------------------Tcold-------------------->Thot
20.0,23.0,25.3,27.3,29.0,30.5,31.8,33.0,34.2,35.2,64.9
20.0,23.0,25.3,27.2,28.9,30.4,31.8,33.0,34.1,35.2,64.7
20.0,23.0,25.3,27.2,28.9,30.4,31.7,32.9,34.1,35.1,64.6
20.0,22.9,25.3,27.2,28.9,30.3,31.7,32.9,34.0,35.1,64.4
19.9,22.9,25.2,27.2,28.8,30.3,31.6,32.9,34.0,35.0,64.2
```

```
19.9,22.9,25.2,27.1,28.8,30.3,31.6,32.8,33.9,35.0,64.1
19.9,22.9,25.2,27.1,28.7,30.2,31.6,32.8,33.9,34.9,63.9
19.9,22.8,25.1,27.0,28.7,30.2,31.5,32.7,33.8,34.8,63.6
19.9,22.8,25.1,27.0,28.7,30.1,31.4,32.6,33.8,34.8,63.4
19.8,22.8,25.1,27.0,28.6,30.1,31.4,32.6,33.7,34.7,63.2
variable: nx=10, ny=10, Q=46.08, Tho=63.15, Tco=34.97
constant: nx=10, ny=10, Q=47.45, Tho=62.84, Tco=34.66
```

Summary

These illustrations reveal the importance of accounting for variations in properties and heat transfer coefficients and also that the analytical methods are not able to handle this. Specifically, averaging the variable parameters and using these in the analytical solutions, hoping to achieve more accuracy is not a rigorous approach because all averaging schemes necessarily presume some distribution throughout the domain and the distributions have been shown (by the preceding contour graphs of temperature differences) to vary from case-to-case, invalidating such presumptions. The numerical approach (implicit FDM) has been shown to rapidly converge and is able to handle all such variations throughout the domain; thus, it will be used when variations are expected.

Chapter 3. Expected Variations

The properties that vary throughout a heat exchanger are either 1) thermodynamic or 2) transport. The thermodynamic property of interest is the constant pressure specific heat, C_P, which appears in the heat transfer equations presented in the previous chapter. Transport properties include: density, viscosity, and thermal conductivity. These three impact the heat transfer coefficient, U, which includes convective and/or phase change (in this case boiling) on the inside and outside of the tubes. The tube material and fouling factors complete the calculation of U.

Specific Heat

The following figure shows the enthalpy of steam (SF-95, see Appendix A) vs. temperature for a series of pressures spanning the range found in a HRSG (atmospheric to critical).

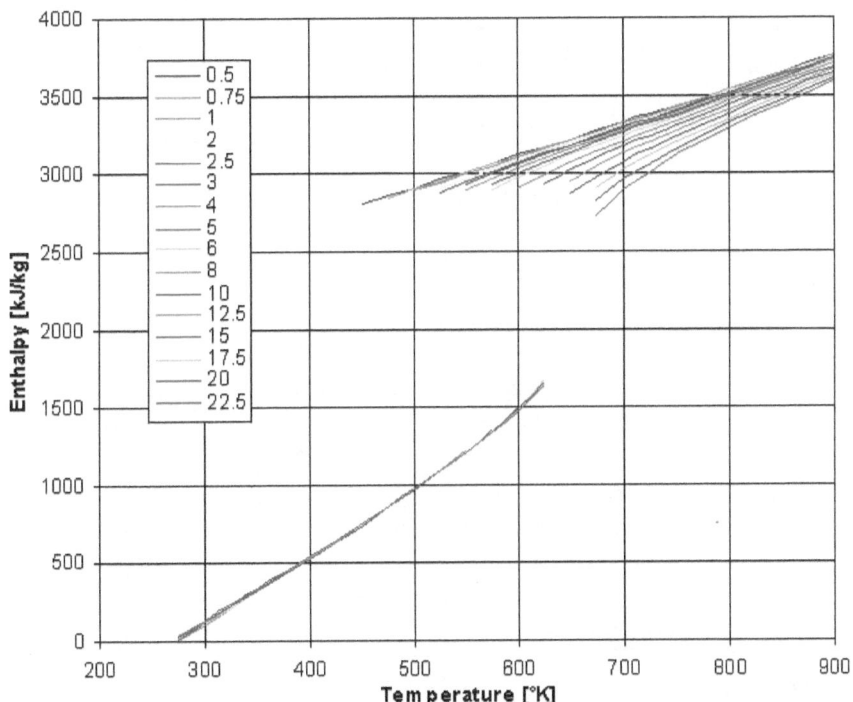

The subcooled liquids are on the lower left and the superheated vapors are on the upper right. All of the subcooled liquids are essentially on top of each other and form an almost straight line, indicating a nearly constant specific heat. This means that using the average specific heat is adequate for any economizer. The superheated vapors do fan out and have different slopes, but very little

21

curvature. This means that using the average specific heat may also be adequate for most any superheater.

We must also consider the gas specific heat, for which we use NASA Glenn (see Appendix B). Typical mole fractions are unfired: 69.90% N2, 11.10% O2, 3.80% CO2, 14.30% H2O, 0.90% Ar and fired: 69.54% N2, 9.59% O2, 4.51% CO2, 15.46% H2O, 0.90% Ar. The mass-weighted specific heats are shown below:

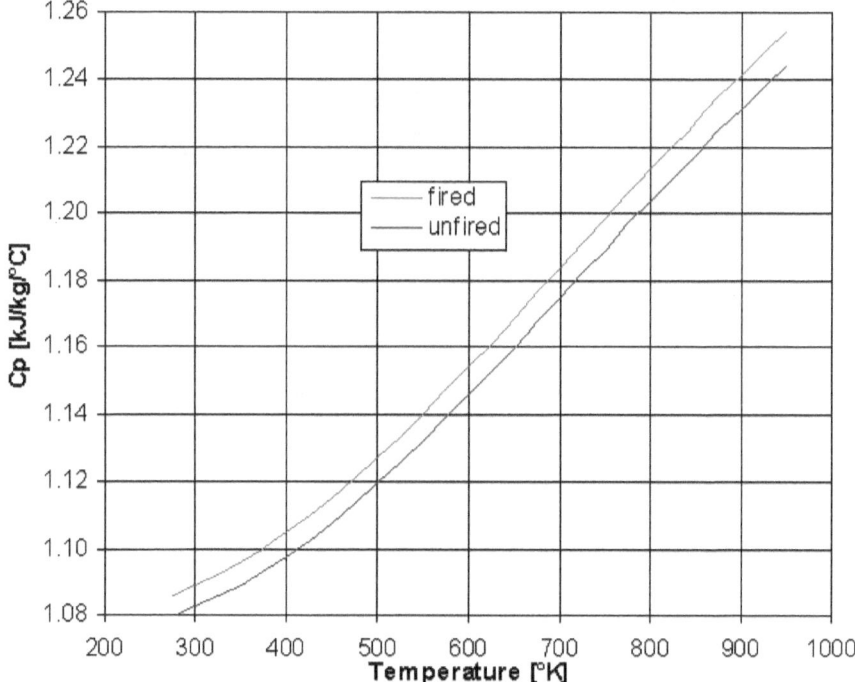

These vary little (<15%) over the range of expected temperatures and do so almost linearly; thus, using the average value will probably be adequate.

Heat Transfer Coefficient

The overall heat transfer coefficient, U, in a typical HRSG will vary at most from 10 to 50 W/m²/°C (2 to 8 BTU/hr/ft²/°F). The lower value might be seen in a significantly fouled or corroded unit, while the higher value would be difficult to achieve in a pristine one. The default value for all economizers, evaporators, and superheaters in GateCycle™ is 8 BTU/hr/ft²/°F. A more reasonable span would be 15 to 25 W/m²/°C (2.5 to 4.4 BTU/hr/ft²/°F). I would be skeptical of reported values outside this range.

The overall heat transfer coefficient is a conductance and accumulates by a series of reciprocals, as in the following equation:

22

$$U = \cfrac{1}{\cfrac{1}{h_{FI}} + \cfrac{1}{h_{CI}} + \cfrac{1}{h_W} + \cfrac{1}{h_{CO}} + \cfrac{1}{h_{FO}}} \qquad (3.1)$$

where h_{FI} is the fouling conductance on the inside of the tubes, h_{CI} is the convection (or evaporation) conductance on the inside of the tubes, h_W is the tube wall conductance, h_{CO} is the convection on the outside of the tubes, and h_{FO} is the fouling conductance on the outside of the tubes. The fouling contributions are most often presented as resistances rather than conductances, that is:

$$r_{FI} = \frac{1}{h_{FI}} \qquad (3.2)$$

Combining the reciprocals in this way results in the smallest value controlling the overall result. The tube wall resistance is quite low (or the conductance is quite high) compared to either convective or evaporative contributions. The convective coefficient is at least an order of magnitude less than the evaporative, making them (inside and/or outside) controlling. As water (and steam) have much higher thermal conductivity and density than exhaust gas, the outside convective coefficient is significantly smaller than the inside, even with fins. See Appendix E for more on heat transfer coefficients.

These considerations lead to the observation that the overall conductance for this type of heat exchanger will most likely be within the range of 15 to 25 W/m²/°C (2.5 to 4.4 BTU/hr/ft²/°F), but not necessarily vary by this much inside a single HRSG component. The biggest differences are specific to the particular design, rather than the location along one tube or another. Entrance and exit effects, as well as headers and manifolds, impact the local conductance, but considerable effort is invested in minimizing such impacts.

Chapter 4. Actual Designs

Reputable HRSG manufacturers know what they're doing. Equipment of this type is very expensive and comes with various guarantees, which if not met, result in significant monetary penalties. The most experienced manufacturers have devoted years to developing and fine-tuning their calculations so as to meet expectations as efficiently as possible. Not surprisingly, they are very protective of this information.

HRSG Design Data

In the folder examples\hrsg of the online archive, you will find a data file (cases.h) and spreadsheet (cases.xls) containing 421 cases from 4 different manufacturers, sufficiently "scrubbed" to obfuscate the source. These cases contain only the thermal performance data and not the proprietary specifics (e.g., tube size, tube length, tube spacing, fin size, fin shape, fin spacing, bending angles, manifold and plenum geometries, etc.). I was personally involved with testing each of these units and know that they performed as expected.

The crossflow program (hrsg.c) has been modified to process this data, displaying the results from all three analysis methods (P-NTU, F-LMTD, and implicit FDM). The calculations are performed using English units, as this is the form in which the design data was supplied. The IF-97 steam properties are used; however, you can easily change this to one of the others (IF-67, KKHM-69, NBS/NRC-84, or SF-95) by modifying the batch file (_compile.bat). The spreadsheet lists the results for all but the FDM and also includes the calculations. Below is a sample:

	A	B	C	D	E	F	G	H	I	J
1	typical HRSG design data									
2	case	area	mC	mH	Pci	Pco	Tci	Tco	Thi	Tho
3	name	ft²	lb/hr	lb/hr	psia	psia	°F	°F	°F	°F
4	1RHTR3	67,711	430,190	3,307,390	309	298	953	1044	1093	1070
5	1HPSH2	127,004	376,950	3,307,390	1277	1232	866	1052	1070	1024
6	1RHTR2	67,711	430,190	3,307,390	311	309	823	953	1024	992
7	1HPSH1	69,351	376,950	3,307,390	1295	1277	577	866	992	893
8	1RHTR1	51,772	430,190	3,307,390	318	311	683	823	893	858
9	1IPSH2	12,859	71,220	1,103,820	325	318	480	555	601	590
10	1LPSH2	25,718	63,710	2,203,570	56	49	479	581	601	595
11	1HPEC2	264,571	376,950	3,307,390	1304	1295	429	569	593	521
12	1LPSH1	25,718	63,710	2,205,260	57	56	289	479	521	510
13	1IPSH1	12,859	71,220	1,102,130	326	326	425	480	521	512
14	1HPEC1	260,865	376,950	2,691,720	1317	1304	292	429	444	366
15	1IPEC	59,288	108,730	615,670	346	326	290	420	444	350
16	1LPEC2	240,115	549,390	3,307,390	70	57	184	284	390	242
17	1LPEC1	156,811	549,390	3,307,390	78	70	81	184	242	174

Figure 4.1 Typical HRSG Design Data

25

	K	L	M	N	O
1			calculations		
2	Qc	Qh	Q	CpC	CpH
3	MBTU/hr	MBTU/hr	error	BTU/lb/°F	BTU/lb/°F
4	20.9	24.5	16%	0.534	0.322
5	42.4	48.6	14%	0.605	0.319
6	29.5	33.5	13%	0.528	0.317
7	89.4	102.4	14%	0.821	0.313
8	32.0	35.7	11%	0.532	0.308
9	3.3	3.5	7%	0.611	0.289
10	3.2	3.8	17%	0.494	0.289
11	62.7	68.2	8%	1.188	0.287
12	64.7	6.9	162%	5.341	0.283
13	2.7	2.8	3%	0.695	0.284
14	54.2	57.8	6%	1.049	0.275
15	14.9	15.9	7%	1.053	0.275
16	55.6	131.4	81%	1.011	0.268
17	56.5	58.4	3%	0.999	0.260

Figure 4.2 Basic Calculation Results

	P	Q	R	S	T	U
1			P-NTU method			
2	Rc	Rh	Pc	Ph	NTUc	NTUh
3	-	-	-	-	-	-
4	0.22	4.62	0.650	0.164	1.21	0.37
5	0.22	4.63	0.912	0.225	100.00	100.00
6	0.22	4.62	0.647	0.159	1.19	0.34
7	0.30	3.34	0.696	0.239	1.52	0.65
8	0.22	4.46	0.667	0.167	1.28	0.36
9	0.14	7.34	0.620	0.091	1.04	0.16
10	0.05	20.26	0.836	0.049	1.92	0.32
11	0.47	2.12	0.854	0.439	100.00	100.00
12	0.54	1.84	0.819	0.047	100.00	0.05
13	0.16	6.31	0.573	0.094	0.92	0.15
14	0.53	1.87	0.901	0.513	100.00	100.00
15	0.68	1.48	0.844	0.610	100.00	100.00
16	0.63	1.60	0.485	0.718	0.86	100.00
17	0.64	1.57	0.640	0.422	1.73	1.18

Figure 4.3 P-NTU Calculations Part A

	V	W	X	Y	Z
1	P-NTU method				
2	UAc	UAh	Uc	Ua	U
3	MBTU/hr/°F	MBTU/hr/°F	MBTU/hr/ft²/°F	MBTU/hr/ft²/°F	error
4	0.277	0.392	4.09	5.79	34%
5	22.792	105.638	179.46	831.77	129%
6	0.271	0.356	4.00	5.25	27%
7	0.470	0.672	6.77	9.70	36%
8	0.293	0.371	5.65	7.17	24%
9	0.045	0.052	3.53	4.04	13%
10	0.061	0.206	2.35	8.02	109%
11	44.769	94.758	169.21	358.16	72%
12	34.030	0.032	1323.18	1.24	200%
13	0.045	0.048	3.53	3.72	5%
14	39.539	74.081	151.57	283.98	61%
15	11.449	16.907	193.11	285.16	38%
16	0.480	88.756	2.00	369.64	198%
17	0.948	1.011	6.05	6.45	6%

Figure 4.4 P-NTU Calculations Part B

	AA	AB	AC	AD	AE
1	F-LMTD method				
2	F	LMTD	UA	U	U
3	-	°F	MBTU/hr/°F	MBTU/hr/ft²/°F	error
4	0.942	78.1	0.308	4.55	8%
5	0.012	64.5	58.386	459.71	9%
6	0.946	113.0	0.295	4.35	6%
7	0.885	206.6	0.524	7.56	8%
8	0.939	114.6	0.315	6.08	5%
9	0.976	73.4	0.047	3.68	3%
10	0.928	54.6	0.069	2.70	48%
11	0.014	50.6	91.432	345.59	31%
12	0.999	107.8	0.332	12.91	98%
13	0.979	61.1	0.046	3.59	1%
14	0.015	37.0	103.950	398.48	83%
15	0.015	39.3	25.318	427.03	79%
16	0.935	79.6	1.256	5.23	97%
17	0.798	74.1	0.972	6.20	1%

Figure 4.5 F-LMTD Calculations

Several things should stand out from even a cursory inspection of the spreadsheet. The first of these is the heat transfer error (column M), that is, the difference between the cold (steam) and hot (gas) side values. Occasionally this arises from a typographic error, but most of the time it is simply round-off in the temperatures. All four of these manufacturers utilize software to perform the calculations, so there shouldn't be a significant internal error other than round-off. There is at least one manufacturer that still performs all calculations by *hand*, that is, without the aid of a computer program—truly hard to believe in the 21st century![3] The maximum error in heat transfer for these 421 cases is 198%, the minimum is 1%, and the average is 34%. I know from extensive testing experience that most of this discrepancy is in the gas side temperatures. Some, but not all, manufacturers assume a certain heat loss from each component (e.g., 5%), which would explain a positive bias in the percent error, but nothing like what you see in the spreadsheet.

P-NTU Results

The next items to notice in the spreadsheet are the differences between the P-NTU results (i.e., U) for the cold side and hot side. If there were no discrepancy between the hot and cold side heat transfer, these two values should be equal. One of the advantages of having this data in a spreadsheet is that it facilitates sorting on any of the columns. Considering only those cases with no more than 10% error in the heat balance, this leaves only 141 cases (33% of the total). Among these, the discrepancy in U is as high as 200%, with an average of 24%.

F-LMTD Results

The discrepancies in U for the F-LMTD method are also large (up to 100%), with an average of 41%. Again, eliminating all those with a heat transfer discrepancy of more than 10%, reduces the average error to 24%.In this one respect, the two methods (P-NTU and F-LMTD) are similar. This is not surprising, considering the expected variations described in the previous chapter.

Implicit FDM Results

The implicit FDM results are calculated by hrsg.c and written out to iFDM.csv, which has been appended to cases.xls as AF:AG. For the full set of cases, these values of U exhibit discrepancies of up to 100% with an average of 41%. Eliminating all but the cases with no more than 10% heat balance error reduces the average discrepancy to 22%.

Adjusted Data

An interesting exercise would be to use the Excel® Solver() to adjust each of the temperatures (or flows) so as to minimize the heat transfer error. The most likely choice of variable is the hot side inlet temperature, as any of the other

---

[3] I offered to write software for them that would faithfully implement their existing process, but they balked at the suggestion.

temperatures may result in a temperature difference violation of the 2nd Law of Thermodynamics (see spreadsheet cases_adjusted.xls). This is also implemented in a code (adjust.c), which creates an output file (adjusted.csv), which has also been converted to an include file (adjusted.h). This include file can be used with the program (hrsg.c) to create an adjusted version of iFDM.csv, which has been appended to adjusted.xls.

Even after adjusting the either the hot side inlet or hot side exit temperatures to minimize the discrepancy in cold side and hot side heat transfer, the average discrepancy in U is still 10%, 5%, and 10% for the P-NTU, F-LMTD, and iFDM methods, respectively.

Reasonableness

Not only must the cold and hot side heat transfer agree, but also the result (U) must be reasonable. The most extreme criterion ($1 \leq U \leq 10$) is met by only 346 of the 421 cases (82% of the total) P-NTU method, 359 for the F-LMTD and 397 for the iFDM. A much more reasonable expectation would be $2 \leq U \leq 8$, which leaves 254, 246, and 278 for the three methods, respectively. The following figure shows all of the U values for all of the cases:

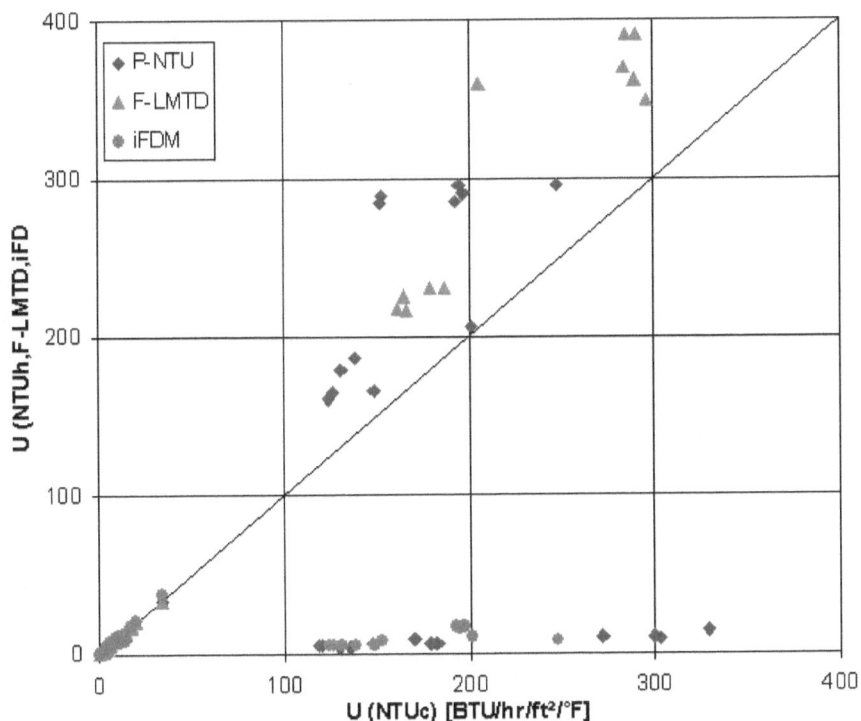

Clearly, many of these are outliers, in spite of having adjusted the gas temperatures to close the heat transfer discrepancies. The bottom left corner of this same figure is shown below:

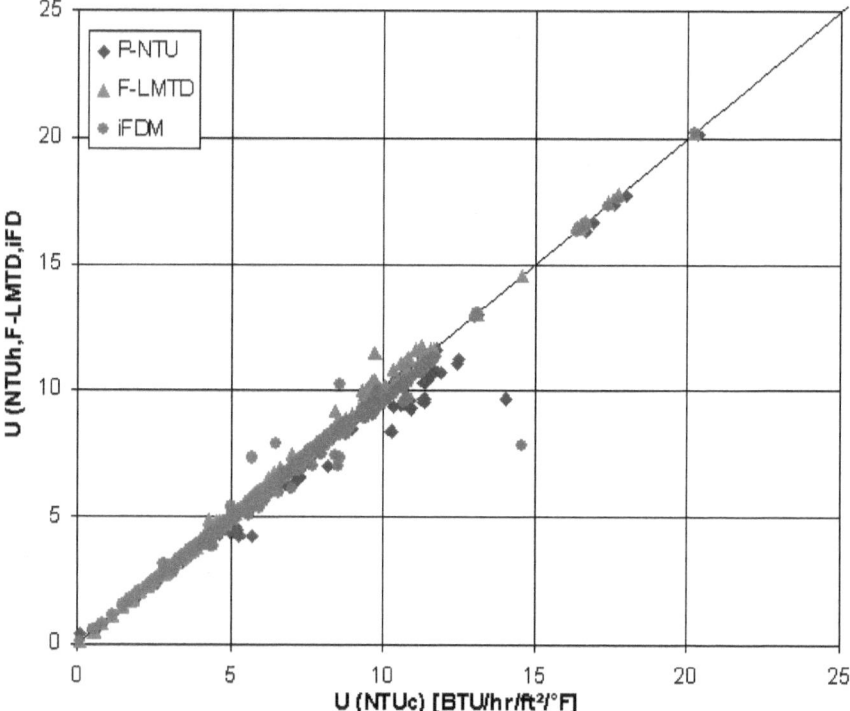

For the somewhat reasonable values ($1 \leq U \leq 12$), all three methods are essentially the same (i.e., the blue diamonds, red triangles, and green circles are practically on top of each other). What this means is that, for this particular type of heat exchanger, it doesn't matter which method you use, as long as the input values are reasonable.

Chapter 5. Heat Release Diagram

Perhaps the most revealing presentation of a HRSG thermal design is the heat release diagram. This is a plot of gas (hot side) and steam (cold side) temperatures vs. the cumulative heat transfer. GateCycle™ produces these graphs for you, but they are not complicated to generate. You will find a spreadsheet (HRSG1.xls) in the online archive in folder examples\release that solves a single-pressure HRSG and generates the associated heat release diagram. There is an English and an SI tab so that it works with both units. Typical results are as follows:

A	B	C	D	E	F	G	H	I
Single Pressure HRSG								
case	mC	mH	Pci	Pco	Tci	Tco	Thi	Tho
name	lb/hr	lb/hr	psia	psia	°F	°F	°F	°F
SH	300,000	3,000,000	1800	1750	621	1050	1100	986
EV	300,000	3,000,000	1800	1800	621	621	986	824
EC	300,000	3,000,000	1950	1800	120	616	824	641

J	K	L	M	N	O	P	Q
Single Pressure HRSG							
Q	CpC	CpH	Rc	Rh	Pc	Ph	NTUc
MBTU/hr	BTU/lb/°F	BTU/lb/°F	-	-	-	-	-
108.86	0.846	0.319	0.27	3.77	0.896	0.237	####
150.73	∞	0.310	∞	0.00	0.000	0.443	0.59
164.13	1.103	0.299	0.37	2.71	0.704	0.260	1.69

R	S	T	U	V	W	X
NTUh	LMTD	F	UA	ΣQ	Th	Tc
-	°F	-	MBTU/hr/°F	MBTU/hr	°F	°F
0.91	158.50	0.78	0.69	0.00	1100	1050
0.59	276.46	1.00	0.55	108.86	986	621
0.60	341.30	0.90	0.48	259.58	824	616
		total	1.71	423.71	641	120
user inputs in blue						
calculations in orange						

The calculations are updated automatically when you change the values in bold blue text.

This next figure is the heat release diagram:

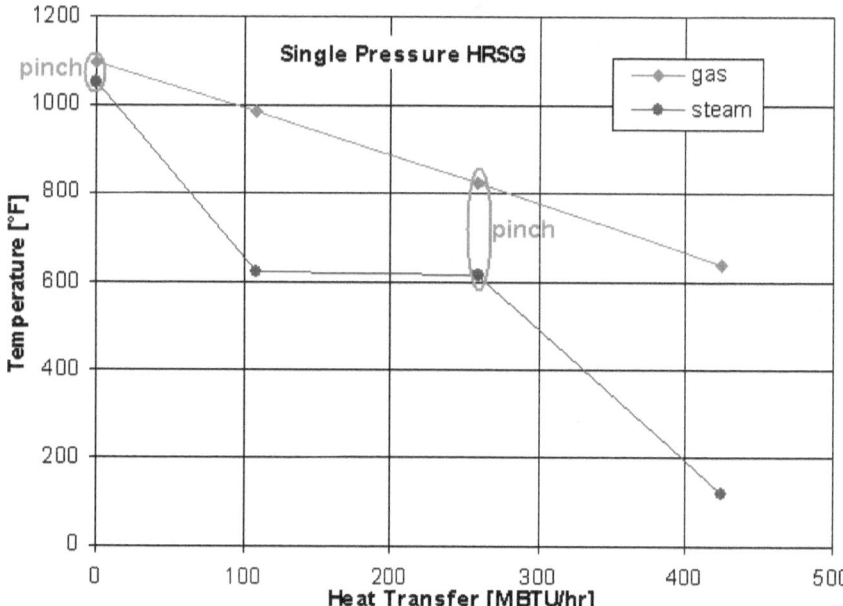

The slope of the lines is equal to minus the reciprocal of the mass flow rate times the specific heat (i.e., $-1/m_C/C_{PC}$ and $-1/m_H/C_{PH}$). There is almost no curvature in the red (hot gas side) line because the specific heat is almost constant over this temperature range. The three sections of the blue line (from left to right) correspond to the superheater, evaporator, and economizer, respectively. The first and third sections of the blue line has negligible curvature, because the specific heat of the vapor and liquid, respectively, is almost constant over the temperature range. The flat portion of the blue line arises from the evaporation process, which is isothermal. The specific heat would be equal to the latent heat of vaporization divided by the temperature change, which is zero, yielding an infinite result. The reciprocal results in a zero slope for this section. While there are some differences from one design to another and single to double to triple pressure systems, the important features of the shape are the same.

<u>Pinch Points</u>

The two locations along the process in the preceding figure where the red and blue lines are closest together are called *pinch* points. The left pinch point occurs at the exit of the superheater. The right pinch point occurs at the exit of the economizer. Pinch points control the operation of a HRSG and are the predominant factor in a thermal design. The required surface area is inversely proportional to the thickness (in degrees) of the pinch points. The tighter the pinch points, the more surface is needed. More surface means more material and more expense.

This maxim is worthy of repeating...
Heat exchangers are sold by the pound (or kg).

To obtain a cost-effective design (and win the bid), the engineer must effectively manage pinch points in a HRSG. The following does not occur:

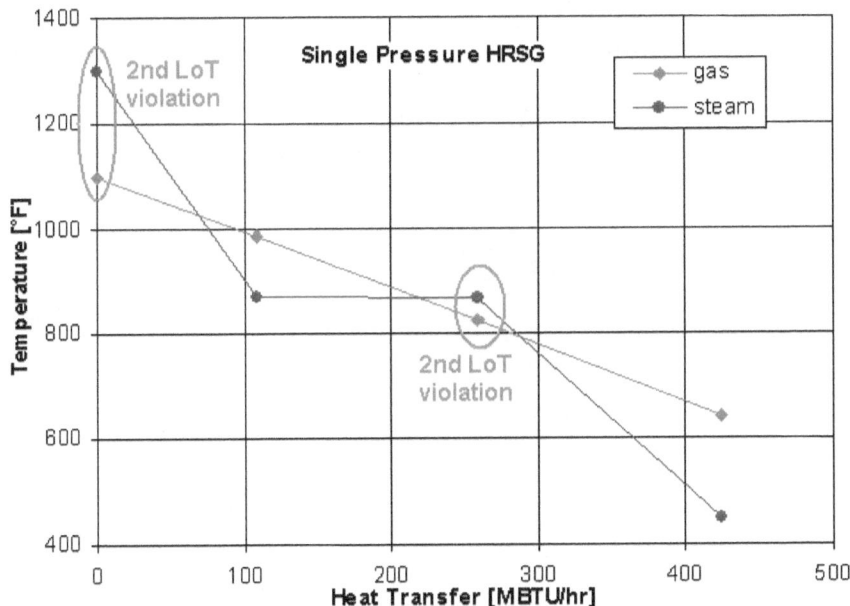

as this would be a violation of the 2nd Law of Thermodynamics (2LoT). Even an infinitely large heat exchanger (A=∞) with infinite conductance (U=∞, or zero thermal resistance) could not achieve this contradiction.[4]

---

[4] Just in case you're interested... Yes, I've seen it on drawings. It's usually a mistake that can be resolved. Twice in 20 years of experience it was a financial disaster (i.e., a lawsuit).

Impact of Steam Flow

There is a button in the spreadsheet to vary the steam flow and produce a modified heat release diagram for three different steam flow rates. There is hardly any impact on the hot side (gas) temperature line except for the end point. The terminal point of the cold side (steam) remains the same also. The flat portion on the steam line remains at the same level, only shifting left to right and growing in length. The right side of the flat zone is closest to the gas line, so steam flow directly impacts the pinch point; thus, the pinch point determines the steam flow.

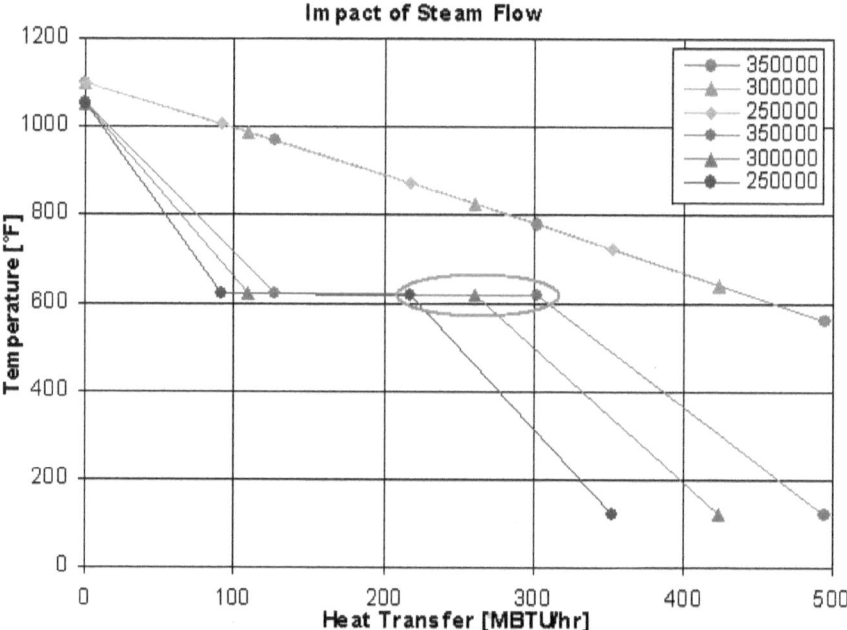

Impact of Steam Pressure

There is a second button that varies the steam pressure. Not only does changing the pressure impact the vertical position (i.e., saturation temperature), but also the length of the flat portion, which changes the position of the end point and reduces the pinch at constant steam flow.

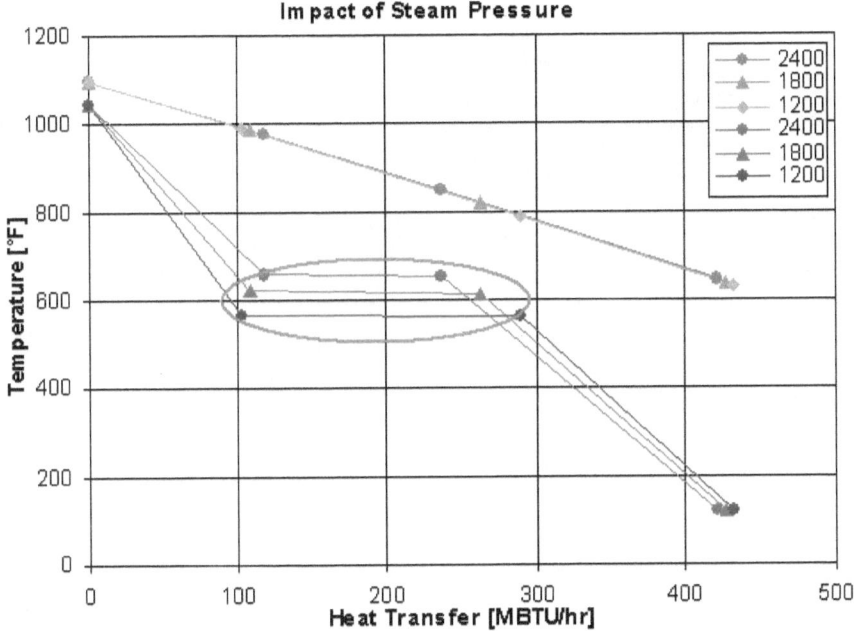

Impact of Gas Flow

The third button varies the gas flow. The slope of the lines (except in the evaporator) is proportional to the inverse of the product of the mass flow and specific heat; so this will change the slope of the hot side (gas) line and also impact the pinch point.

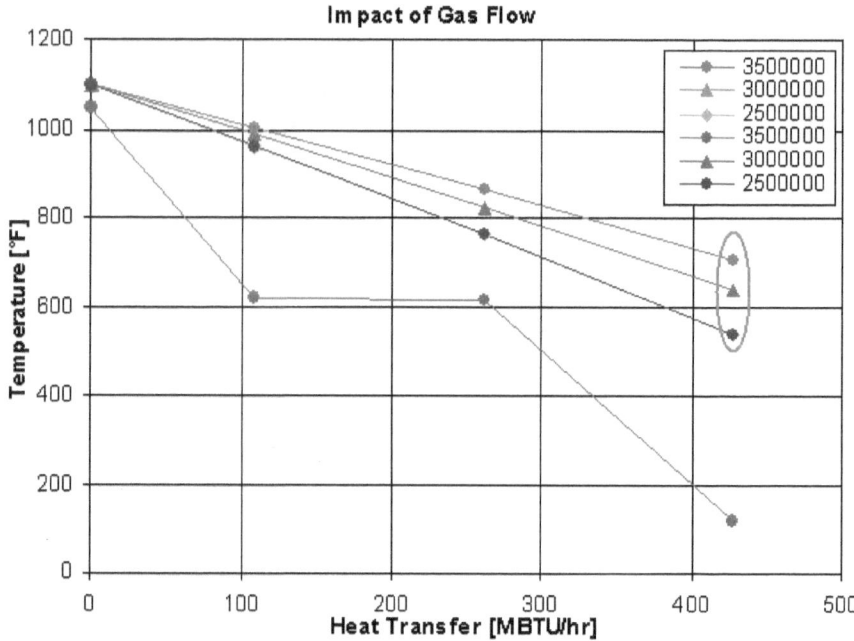

Impact of Gas Temperature

The fourth button varies the gas temperature. The slope of the hot (gas) line doesn't change, but the whole line shifts up and down; so this will impact the pinch point.

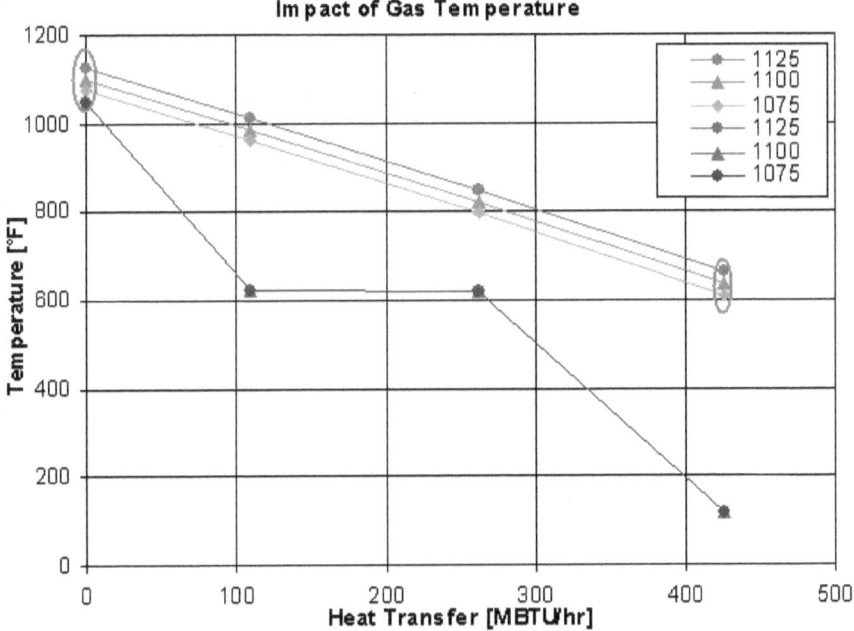

Three-Pressure HRSG

The most common arrangement is a three-pressure HRSG. The low pressure (LP) is mostly for control, venting, and purification. The intermediate pressure (IP) is added to the cold reheat and the high pressure (HP) forms the main steam. The reheat is necessary because continued expansion of the steam without reheating would result in far too much moisture and damage of the steam turbine. This is readily apparent from a Mollier Diagram. A typical GateCycle™ heat release diagram for a three-pressure system is shown in this next figure:

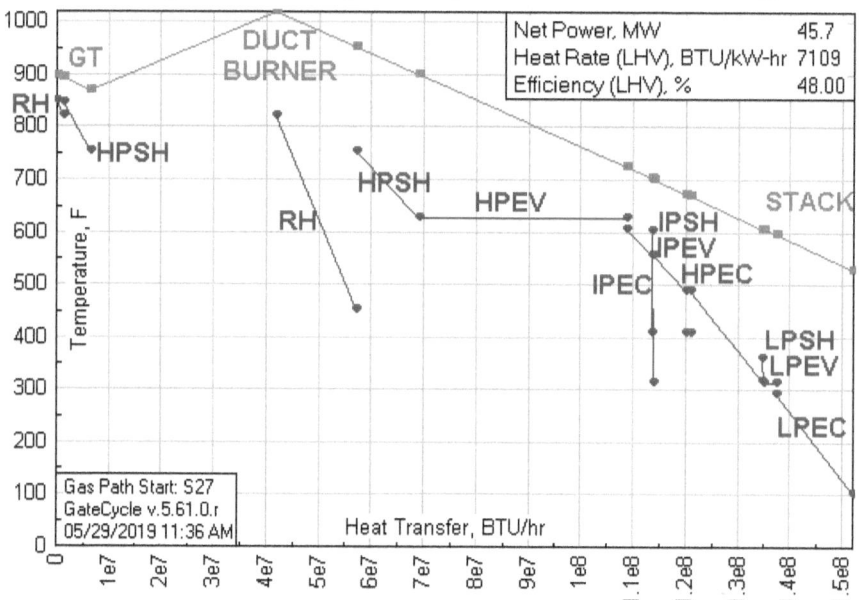

Impact of Duct Firing

Notice in the above figure how the duct burner pushes the red (hot gas) side temperature up to the right and also moves the right-hand-side farther away from the blue (cold steam) line, relieving the pinch. This is why duct firing is so prevalent in combined cycle power plants (CCPPs). While it's beyond the scope of this book, the incremental heat rate for duct firing is not what you might expect. The overall plant test code, ASME PTC46[5], there is a correction for duct firing, known as ω_7/Δ_7. The change in heat input (from design or guarantee point) is ω_7 and the associated change in net power output is Δ_7. While the correction is applied downward (more heat input from the duct burner requires

---

[5] *Performance Test Code on Overall Plant Performance*, American Society of Mechanical Engineers, 2003.

an adjustment downward in net power output to arrive at the corrected value), the graph is most often drawn with an upward slope, as illustrated below:

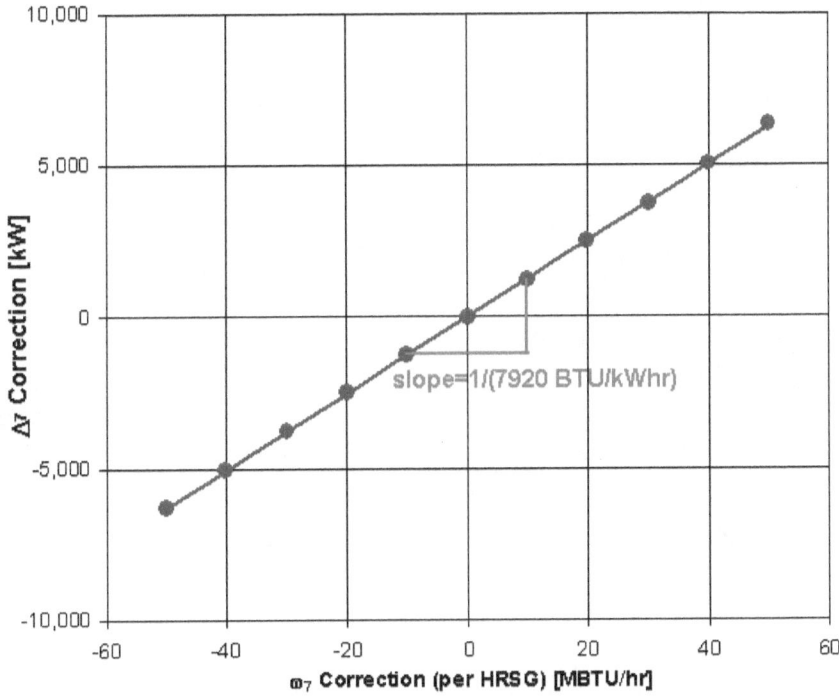

Two things about this figure are quite interesting: 1) it's a straight line and 2) the slope is less (worse) than the net plant heat rate, which in this case is 6650 BTU/kWhr. Duct firing is used to produce more power, but it isn't *cheap* power, because it increases the heat rate (decreases the efficiency). I have built well over 100 GateCycle™ models. More than 50 of these have been complete, including correction curves. In all of my experience, I have never seen a case in which the slope of the ω_7/Δ_7 curve was equal to the reciprocal of the base net plant heat rate, nor have I ever seen a case with significant curvature to this line.

Typical Three-Pressure CCPP

We will discuss the placement of elements in the HRSG in a subsequent chapter. For now, we will work with a typical design and see how to build a spreadsheet to perform the calculations in the absence of commercial software. You will find a an example along with the source code (CCPP4.c) in the online archive that accompanies my text, *Thermodynamic Cycles*. We will not dwell on how to solve the system or attach the properties, as we are only considering the thermal design of HRSGs in this present text. The current example system is illustrated in the following figure:

39

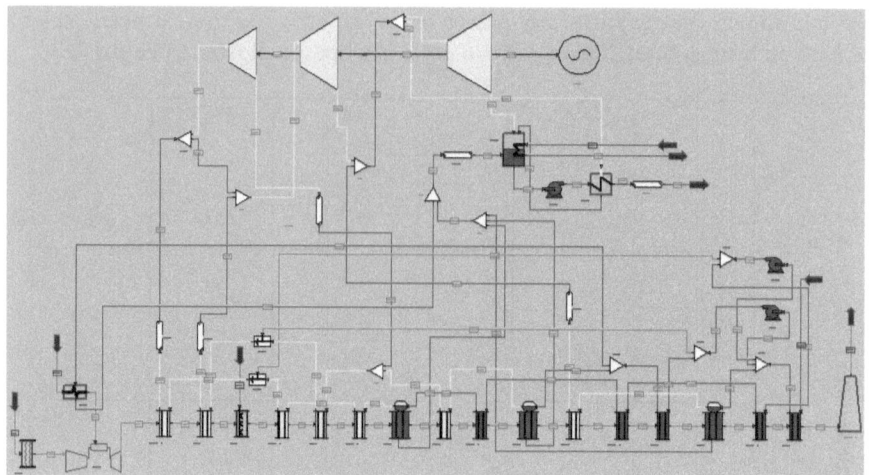

The corresponding spreadsheet (CCPP3.xls) is in folder examples\release, which only contains the HRSG. The entire GateCycle™ model can be found in the folder examples\CCPP. The heat release diagram for this system is shown below. GateCycle™ displays all of the steam paths in the same color and as disjointed segments, which is understandable but may be confusing. The Excel® spreadsheet allows us to draw all three steam paths distinctively.

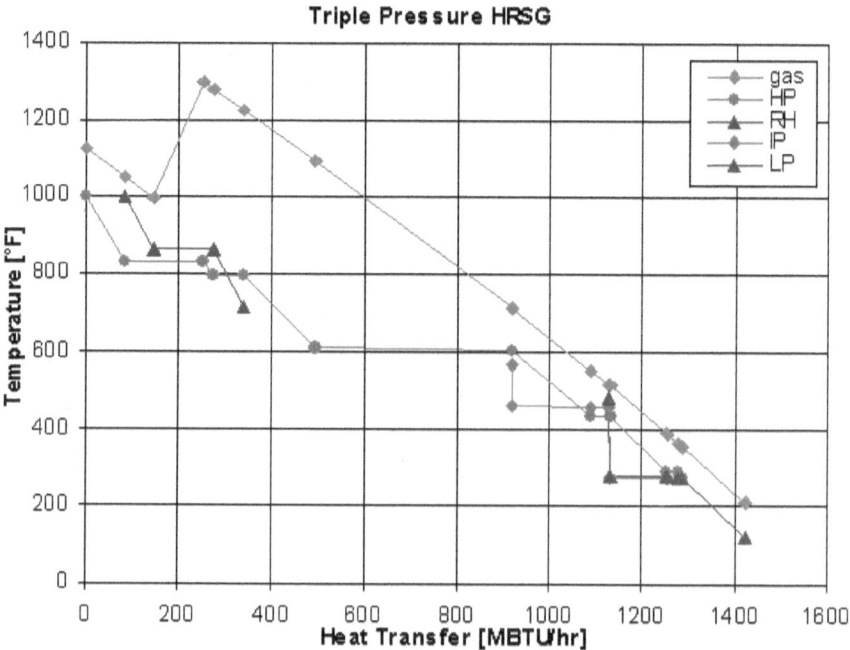

This spreadsheet also as an English and SI tab. The same data in SI units is shown in this next figure:

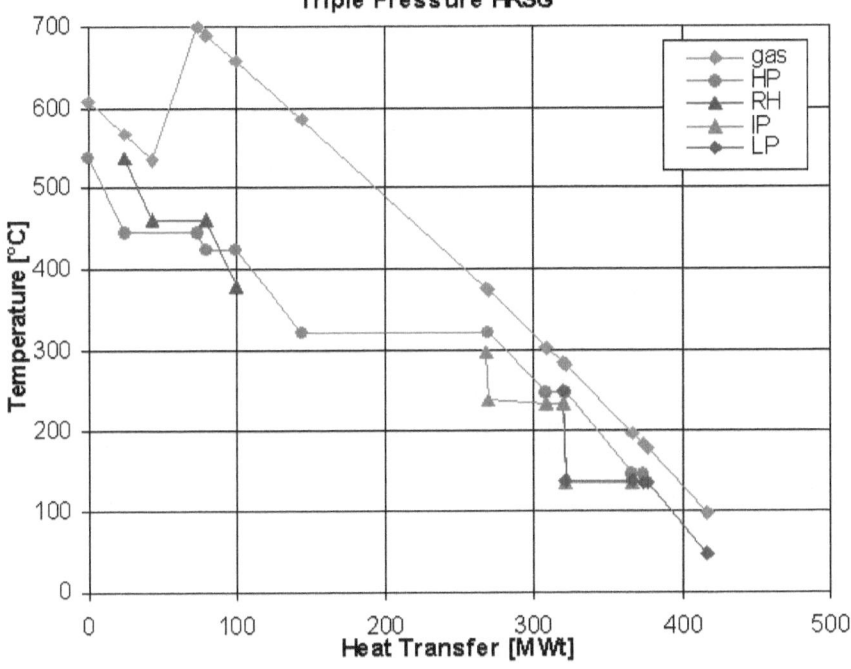

Chapter 6. Expansion Line

The Mollier diagram shows enthalpy vs. entropy and is an essential graphical representation of the expansion process that occurs in a steam turbine. A complete Mollier diagram can be found in Appendix A. In this chapter we will only consider part of the total figure, that is, the expansion line. The following figure (see expansion.xls in the examples\expansion folder) shows a typical expansion line:

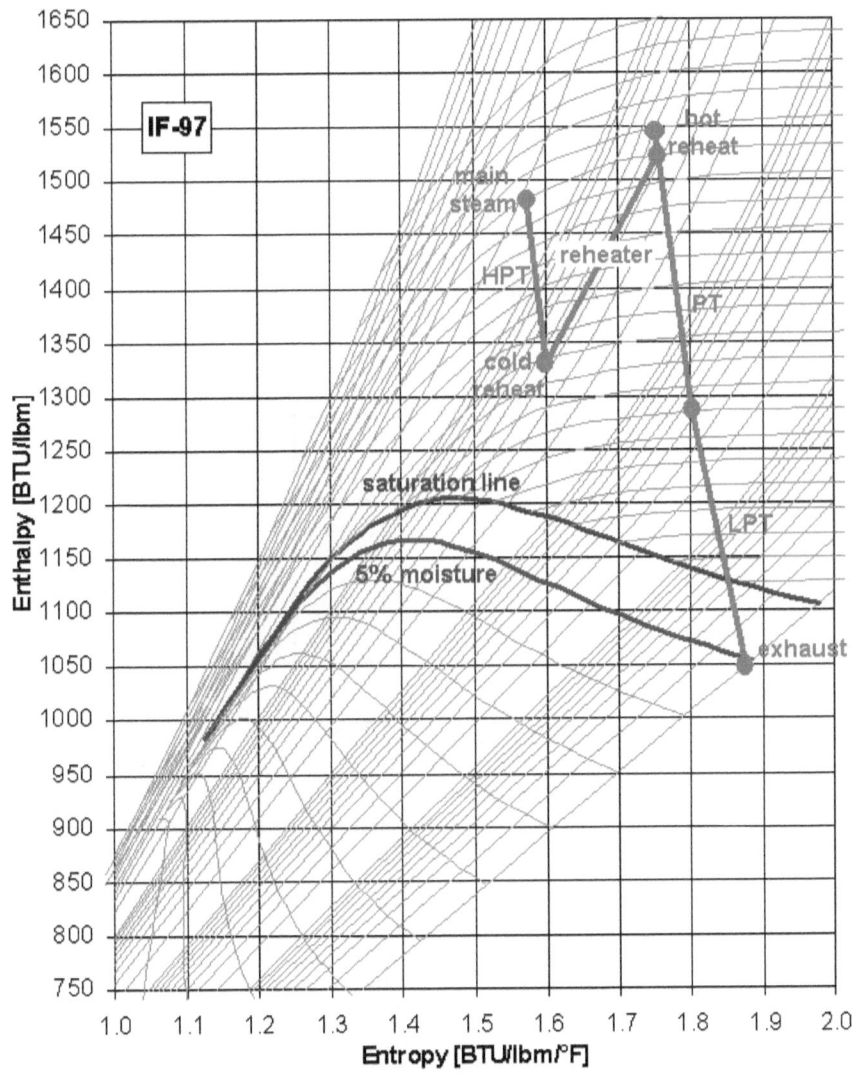

3

The reason a Rankine cycle uses reheat is to avoid dipping down too far into the saturation zone (below the brown saturation line). The power output is proportional to the total vertical drop of the red expansion line, so we want that to be as long as possible. Reheat moves the HRH point up along the isobar allowing for more drop before ending at the 5% moisture line in this case. If you want higher efficiency (power output divided by heat input), reheat is necessary. The practical HRSG must provide this.

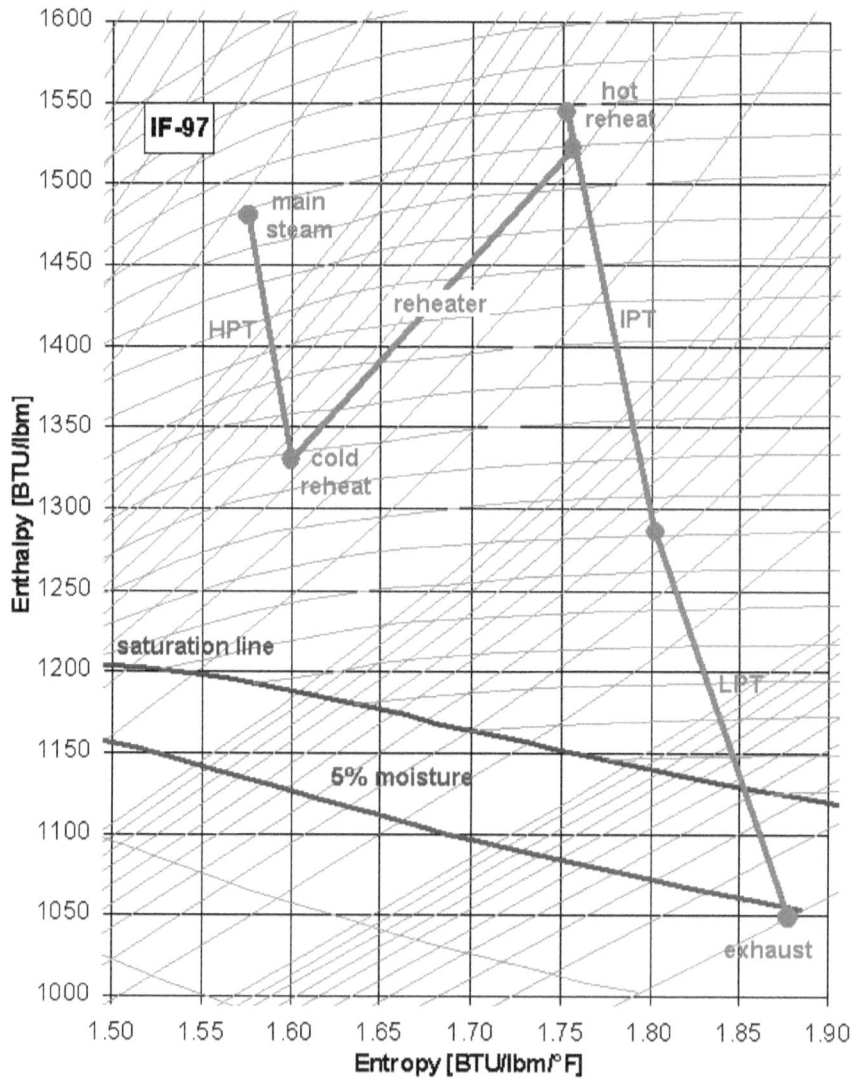

Notice that the HPT (high pressure turbine) drops down to the CRH (cold reheat) and may not come close to the saturation line. The IPT (intermediate pressure turbine) and LPT (low pressure turbine) expansion dips through the saturation line. The length of the HPT expansion (150.7 BTU/lbm) is significantly shorter (70% less) than the IPT+LPT expansion (495.3 BTU/lbm). It is often assumed that the high-pressure turbine does more work than the low-pressure turbine but the opposite is true. Steam turbines are volumetric expansion devices. The really big one (LPT) does most of the work. The little one (HPT) just sets things up for the LPT. The whole point of reheating the steam is to keep it out of the wet region (>5% moisture).

As illustrated in this figure, the CRH is still quite superheated (192°F). The exhaust is where you need to be concerned for moisture. This is why you will sometimes see the first heat exchange element (superheater) in a HRSG at the exit of the GT exhaust (and duct burner exit) is a reheater instead of a main steam one. If you must choose between higher HRH and higher MS (main steam) temperatures, the HRH is more critical in many systems, depending on the pressures and the steam turbine specifics. Again, it is often assumed that hotter main steam means more power output, but hotter reheat steam is more valuable in terms of available expansion line length and specific power output (enthalpy drop).

By the way... I have been asked more than once, "Why bother condensing the steam? Why not just run it through again and not waste all that heat?" Recall that the steam turbine is a volume expansion device? Well try compressing *steam* back to 1800 psia (or 12 MPa)! You really don't want to do that, as it will take more power than you got out of the turbine when expanding it. The differential work for an open system (e.g., a steam turbine or boiler feed pump), the differential work is:

$$dW = VdP = \frac{dP}{\rho} \qquad (E.1)$$

and *that's* why you want to expand vapors and compress liquids! Go to a power plant and look at the size of the components and the diameter of the shafts: low-pressure steam turbine vs. boiler feed pump (BFP). An LPT can be the size of a school bus and the corresponding BFP the size of a dishwasher.

Chapter 7. Element Arrangement

Optimal arrangement of elements means the most heat transfer with the least surface area and resulting material cost. This is accomplished by positioning the segments of the cold (steam) side process lines in the heat release diagram so that the distance between these and the hot (gas) side process line is fairly uniform over much of the range without creating pinch points. Of course, the width of the evaporator (flat) section is unavoidable, given the operating pressures (determined by the steam turbine swallowing capacity). We could view this like a tangram:

in the sense that we can move the sections around in order to obtain the desired shape, recognizing that there are a limited number of solutions and possible shapes.

The geometric shapes represented by the line segments in the heat release diagram are no more complicated than the ones above. We will use the three-pressure example (CCPP3.xls) to illustrate this concept. First, consider just the process lines:

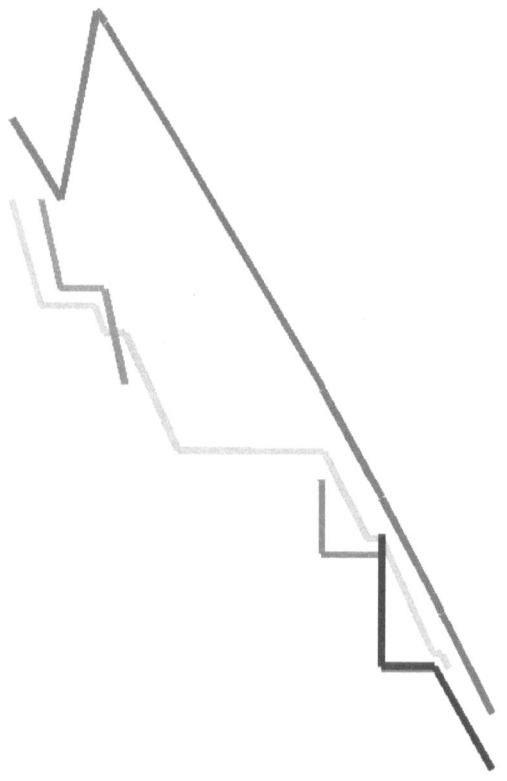

The horizontal shifts other than in the evaporator sections arise from having an intervening heat exchanger along a different path (HP, RH, IP, or LP). We can collapse these gaps in order to illustrate a point. Of course, the segments are no longer properly aligned and don't represent the actual pinch point(s).

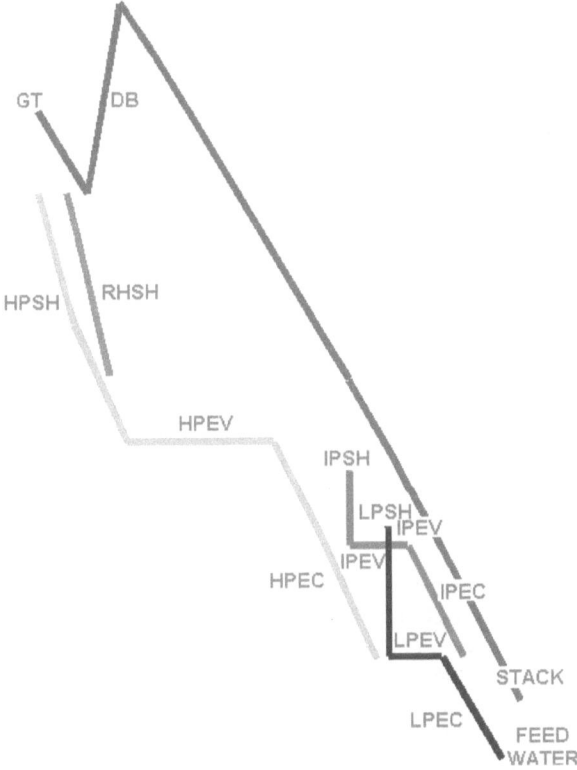

Except for the reheater, which has no evaporator section, the process lines all have the same shape: rise, flat, rise again at a different slope. The slope of each section is equal to minus the reciprocal of the mass flow rate times the specific heat. It doesn't matter how many economizer or superheater sections you have along a process line, the shape will be the same. The only difference will be the horizontal shifts (Q-axis discontinuities) representing the intervening components.

The more sections you split the HP economizer into (or the superheaters), the more fine-tuning available to minimize the pinch points. There is no obvious upper limit from a thermal point of view, but there certainly is from a mechanical and economic one. There's pressure drop with each manifold and from one section to another. Pressure drop generates entropy, reduces free energy (i.e., available for doing work), and is wasteful. Too many components increases expense and is not cost-effective.

49

Split Components

You will often see split components in a HRSG. There are several among the 421 examples in spreadsheet cases.xls. In such cases, the total gas flow is split into two streams—at least conceptually, for there may or may not be an actual baffle in the duct. This need not be an even 50/50 split. The following system has an absurd number of split components:

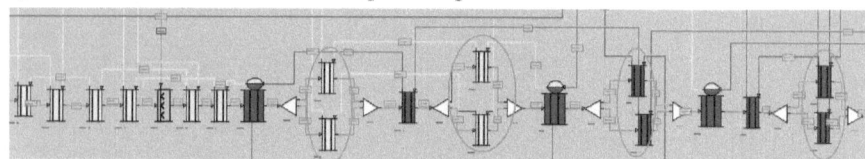

All three types of components (economizer, evaporator, and superheater) can appear in one of these splits. I have never seen a three-way split, although it is certainly possible, at least physically. Splits like this only make sense from a thermal perspective when the normal operating IP and LP steam flows are very much smaller than the HP steam flow. Even when this does occur, it's because the manufacturer wants to maintain a minimum number of rows of tubes in the direction of gas flow. This strategy can also make some sense from a mechanical or economic point of view if the manufacturer has only a limited number of designs for each type of element, fabricates them in advance, or obtained them from some other project. This latter situation is often called buying cheap stuff on the *gray market*.

You will almost never see the same exit gas temperature from two parallel components. The difference may be a few degrees up to tens. Any time you mix two different temperatures, entropy is generated—and cannot be reversed. When entropy is increased, the free energy (g=h-Ts) is reduced along with the available work. This does not make sense thermodynamically. Use split components if you must, but don't say, "Carnot made me do it!"[6]

First HP Economizer

How far down the gas side do you position the first HP economizer? That's a good question. Occasionally, you will see a design with an HP or IP economizer element downstream of the LP evaporator. CCPP3.xls contains one of these designs, which I put there for this very reason. The order (from GT to stack) is: HPSH1, RH1, DB, HPSH2, RH2, CRH, HPSH3, HPEV, IPSH, HPEC1, IPEV, LPSH, HPEC2, IPEC, LPEV, HPEC3, LPEC. Something you should notice is that the LP steam flow will be much lower (even zero) for duct-fired operation than without duct firing. This is all driven by the shifting pinch

---

[6] Carnot refers to the theoretical maximum efficiency one can achieve with a heat engine operating between two temperatures. Named after Nicolas Léonard Sadi Carnot (1796–1832) French military scientist and physicist.

points caused by the bump up of the gas temperature line, as illustrated previously.

You must design the HRSG so that it will operate properly with and without duct firing. Sometimes there a trade-off is necessary to obtain an acceptable solution to both operating conditions. The first (from the feed water end) HP economizer component can shift the pinch point, avoiding excessive stack temperatures, which are wasted heat. You should also notice that the stack temperature will be lower for fired than unfired operation—again due to the shifting pinch point and trade-off of designs.

Optimization

It would be rather difficult to swap around the order and number of components in an Excel® spreadsheet such as CCPP3.xls. However, this is an easy task in the C programming language and we already have most of the code necessary. There might be something learned from swapping components around randomly, but not all combinations make sense. While I have seen at least one case where an HP economizer was upstream of an HP evaporator, this would not likely be an optimal solution. I have also seen designs where the HP evaporator is broken up into several sections and positioned between one or more superheaters. Once we develop a general code, we could investigate the merits of such an arrangement. You will find the code (CCPP3.c) in the folder examples\arrange. There are also two other similar programs that solve a combined cycle power plant, CCPP1.c and CCPP4.c in the folder examples\CCPP. The first uses SI units and the second uses English. The HRSG data statements in CCPP3.c can be modified and rearranged to handle other configuration, although there are limitations, which the code checks. The elements are defined in a data statement:

```
HRSG hrsg[]={
/*   name      st ty */
  {"HPSH1a",HP,SH},
  {"HPSH1b",HP,SH},
  {"RH1a"  ,RH,SH},
  {"RH1b"  ,RH,SH},
  {"DB"    ,NA,DB},
  {"HPSH2a",HP,SH},
  {"HPSH2b",HP,SH},
  {"RH2a"  ,RH,SH},
  {"RH2b"  ,RH,SH},
  {"CRH"   ,NA,CR},
  {"HPSH3a",HP,SH},
  {"HPSH3b",HP,SH},
  {"HPEV"  ,HP,EV},
  {"IPSHa" ,IP,SH},
  {"IPSHb" ,IP,SH},
  {"HPEC1a",HP,EC},
```

51

```
{"HPEC1b",HP,EC},
{"IPEV"  ,IP,EV},
{"LPSHa" ,LP,SH},
{"LPSHb" ,LP,SH},
{"HPEC2a",HP,EC},
{"HPEC2b",HP,EC},
{"IPECa" ,IP,EC},
{"IPECb" ,IP,EC},
{"LPEV"  ,LP,EV},
{"HPEC3a",HP,EC},
{"HPEC3b",HP,EC},
{"LPECa" ,LP,EC},
{"LPECb" ,LP,EC}};
```

The stream can be any one of: HP (high pressure), RH (reheat), IP (intermediate pressure), LP (low pressure), or NA (not applicable). The type can be any one of: SH (superheater), EV (evaporator), EC (economizer), DB (duct burner), or CR (cold reheat). You can comment out repeated elements, as illustrated above for HPEC1b, HPEC2b, and HPEC3b. You can also add more elements and even swap them around, as illustrated below for indices i and j:

```
void Swap(int i,int j)
  {
  HRSG h;
  if(i!=j)
    {
    h=hrsg[i];
    hrsg[i]=hrsg[j];
    hrsg[j]=h;
    }
  }
```

You can change the gas turbine exhaust flow and temperature (EGW, EGT), the superheat exit temperatures, duct burner temperature, and so forth to investigate various combinations. You could even swap the elements randomly and save each combination that improves the estimated net power output.

Typical program output is listed below:

```
3-pressure HRSG design
results file: arrange.csv
solving HRSG
GT flow=3600000
HPEV flow=780144
HP pressure=2580,1680,1500
IPEV flow=41781
IP pressure=520.0,480.0,460.0
RH pressure=460,452
CRH flow=750498
HRH flow=792279
```

52

LPEV flow=33062
LP pressure=148.00,48.00,46.00
HP temperatures=283,612,1000
6 HP economizers
6 HP superheaters
IP temperatures=279,463,565
2 IP economizers
2 IP superheaters
4 RH superheaters
LP temperatures=120,278,480
2 IP economizers
2 LP superheaters

name	mC	Pci	Pco	Tci	Tco	mH	Thi	Tho
HPSH1a	780144	1530	1500	935	1000	3600000	1125	1098
HPSH1b	780144	1560	1530	871	935	3600000	1098	1069
RH1a	792279	454	452	929	1000	3600000	1069	1043
RH1b	792279	456	454	858	929	3600000	1043	1016
DB	0	0	0	0	0	3614500	1016	1316
HPSH2a	780144	1590	1560	806	871	3614500	1316	1287
HPSH2b	780144	1620	1590	741	806	3614500	1287	1254
RH2a	792279	458	456	787	858	3614500	1254	1229
RH2b	792279	460	458	716	787	3614500	1229	1203
CRH	750498	460	460	725	716	3614500	0	0
HPSH3a	780144	1650	1620	676	741	3614500	1203	1163
HPSH3b	780144	1680	1650	612	676	3614500	1163	1101
HPEV	780144	1680	1680	607	612	3614500	1101	732
IPSHa	41781	470	460	514	565	3614500	732	730
IPSHb	41781	480	470	463	514	3614500	730	729
HPEC1a	780144	1830	1680	553	607	3614500	729	674
HPEC1b	780144	1980	1830	499	553	3614500	674	627
IPEV	41781	480	480	458	463	3614500	627	596
LPSHa	33062	47	46	379	480	3614500	596	595
LPSHb	33062	48	47	278	379	3614500	595	593
HPEC2a	780144	2130	1980	445	499	3614500	593	548
HPEC2b	780144	2280	2130	391	445	3614500	548	504
IPECa	41781	500	480	369	458	3614500	504	500
IPECb	41781	520	500	279	369	3614500	500	496
LPEV	33062	48	48	278	278	3614500	496	466
HPEC3a	780144	2430	2280	337	391	3614500	466	423
HPEC3b	780144	2580	2430	283	337	3614500	423	380
LPECa	854987	98	48	199	278	3614500	380	310
LPECb	854987	148	98	120	199	3614500	310	239

name	NTUc	NTUh	F	LMTD	UAc	UAh	UAf
HPSH1a	0.46	0.19	0.986	142.9	0.225	0.225	0.225
HPSH1b	0.36	0.16	0.991	179.9	0.186	0.186	0.186
RH1a	0.82	0.30	0.962	89.6	0.352	0.350	0.352
RH1b	0.54	0.20	0.983	134.8	0.229	0.228	0.229
HPSH2a	0.14	0.06	0.999	463.3	0.077	0.077	0.077
HPSH2b	0.13	0.07	0.999	497.0	0.079	0.079	0.079

```
RH2a     0.17 0.06 0.998 418.4 0.073 0.073 0.073
RH2b     0.15 0.06 0.999 463.8 0.067 0.067 0.067
HPSH3a   0.14 0.08 0.998 473.9 0.101 0.101 0.101
HPSH3b   0.13 0.13 0.997 488.0 0.148 0.148 0.148
HPEV               1.000 267.3             1.554
IPSHa    0.27 0.01 1.000 190.6 0.008 0.011 0.008
IPSHb    0.21 0.01 1.000 240.5 0.007 0.011 0.007
HPEC1a   0.46 0.46 0.970 122.1 0.493 0.493 0.493
HPEC1b   0.44 0.39 0.974 124.7 0.418 0.418 0.418
IPEV               1.000 150.6             0.212
LPSHa    0.63 0.01 0.999 160.6 0.010 0.010 0.010
LPSHb    0.39 0.01 1.000 261.8 0.007 0.010 0.007
HPEC2a   0.57 0.48 0.961  98.2 0.499 0.499 0.499
HPEC2b   0.52 0.42 0.968 107.5 0.432 0.431 0.432
IPECa    1.11 0.05 0.991  81.3 0.050 0.050 0.050
IPECb    0.52 0.02 0.998 170.4 0.023 0.023 0.023
LPEV               1.000 202.2             0.151
HPEC3a   0.71 0.57 0.943  79.9 0.578 0.577 0.578
HPEC3b   0.62 0.49 0.956  90.7 0.492 0.491 0.492
LPECa    0.81 0.72 0.924 105.8 0.701 0.700 0.701
LPECb    0.74 0.66 0.935 114.7 0.631 0.630 0.631
total HP/RH conductance=6.0
total IP conductance=0.3
total LP conductance=1.5
estimated net power 124637 kW
```

The steam flows are adjusted to match the specified overall conductances and the net power output is estimated based on typical condenser conditions (expansion line end point enthalpy). This facilitates comparing various arrangements.

Cost Effectiveness

One of the first things to consider is estimated net power output divided by material costs (total UA), which is the cost-effectiveness. This answers the question, "How much surface area (material) is best?" To do this, just put the target overall conductances (total UAs) in a loop and let 'er rip! You can write the results to a CSV file and then pull it right into Excel®. One impact we expect to see with increasing UA is decreasing stack temperature, as more energy is being removed from the gas stream into the steam. The stack temperature should asymptotically approach that of the feed water. This next figure illustrates this:

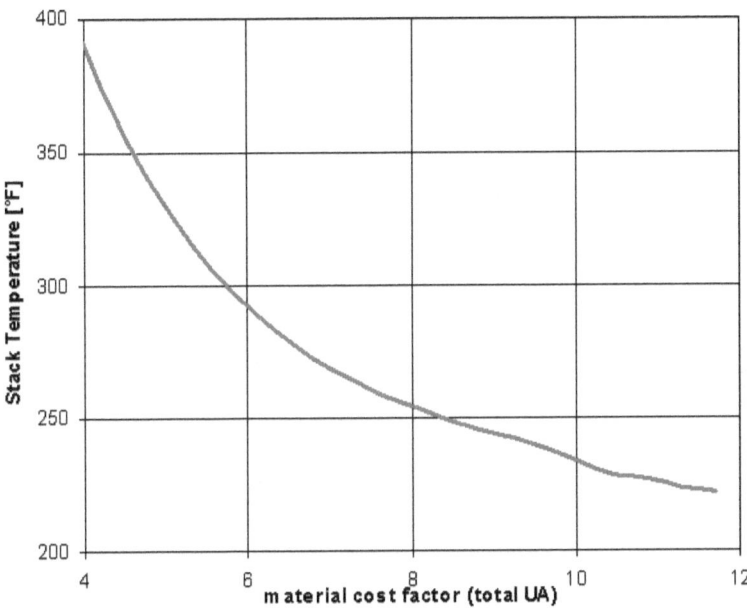

As more steam is produced with greater surface area, the estimated net power output should also increase asymptotically to the level that approaches recovery of the GT exhaust heat less turbine and generator efficiencies.

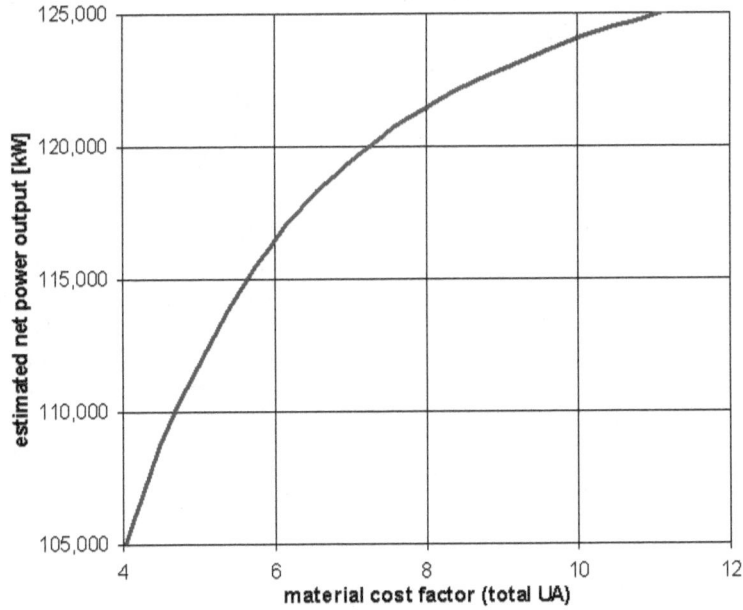

The HRSG is only part of the cost. There is also the steam turbine, generator, pumps, piping, condenser, cooling tower, control system, and much more. Using typical values for such a plant, we obtain the following curve of cost-effectiveness.

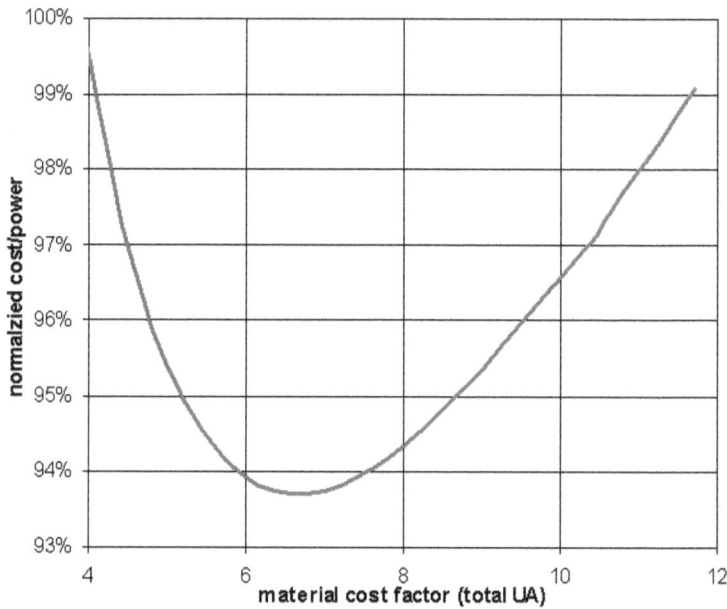

Paying twice as much for a HRSG with twice the UA to obtain 15% more steam flow is not cost-effective. Blowing 400°F/200°C gas into the sky is stupid, considering you could generate electricity, dry wood, heat a building, or bake a lot of pizza.[7] If your curve doesn't exhibit a most cost-effective point, there's something wrong, because this is the way the world works.

Sectionalized Components

One of the actual designs in cases.xls is shown on the next page. This particular design has several elements of the same type and stream immediately adjacent/sequential: LPEC1-4, LPEV1-3, HPEC1-2, IPEV1-3, HPEC4-5, HPEV1-4, and HPSH1-2. This is done for mechanical/physical considerations, not thermal. We can easily investigate using the program (arrange.c) by adding duplicate components.

---

[7] Yes, I've seen it and wondered how many birds would eventually die and what would happen if a helicopter flew over this hot blast on its way to the nearby hospital. It's humorous to point out that this design was cooked up by a professor at a prestigious university who has clearly never worked in industry but likely has many esteemed publications in the annuls of heat transfer.

GT
HPSH4
RH3
HPSH3
RH2
DB
HPSH2
HPSH1
RH1
HPEV4
HPEV3
HPEV2
HPEV1
HPEC6
SCR
IPSH
HPEC5
HPEC4
LPSH
IPEV3
IPEV2
IPEV1
HPEC3
IPEC
HPEC2
HPEC1
LPEV3
LPEV2
LPEV1
LPEC4
LPEC3
LPEC2
LPEC1
STACK

The following table shows the difference in performance resulting from splitting every one of the superheaters and economizers into two equal halves:

splitting superheaters and economizers				
result	units	1 each	2 each	diff.
main steam	lb/hr	757,897	763,261	0.71%
stack temp.	°F	256	240	-16
net power	kW	121,120	122,304	0.98%

Imagine the expense in piping and pressure drop required to get 0.7% more main steam, a 16°F drop in stack temperature, and 1% more net power? Manufacturers do this because they have elements already designed and ready to fabricate of various sizes.

Rearranging Elements

We will now consider rearranging the elements. This is easily accomplished with the code listed previously. The first example will be to swap the first four elements from HPSH1, HPSH2, RH1, RH2 (case 1) to RH1, RH2, HPSH1, HPSH2 (case 2). Then to HPSH1, RH1, HPSH2, RH2 (case 3) and RH1, HPSH1, RH2, HPSH2 (case 4). The results are:

case no.	Fms lb/hr	Tstk °F	power kW
1	763,261	239.8	122,304
2	762,822	239.9	122,283
3	764,885	239.5	122,385
4	765,129	239.4	122,397

The differences are so small as to be negligible. While there is little point swapping around the last main steam and reheat superheater sections, some improvement is possible by swapping around the IP/LP components and the HP economizer elements. We can adapt the code (arrange.c) to randomly swap two components, resolve, and compare to the previous results, keeping track of any improvements. After 2079 combinations, 1001 cases converged and 33 improvements were found. The initial configuration and results were listed 5 pages ago and the most improved case is listed below:

```
solving HRSG
GT flow=3600000
HPEV flow=795928
HP pressure=2580,1680,1500
IPEV flow=38581
IP pressure=520.0,470.0,460.0
RH pressure=460,452
CRH flow=765683
HRH flow=804264
```

58

```
LPEV flow=66254
LP pressure=98.00,48.00,45.00
HP temperatures=283,612,1000
6 HP economizers
6 HP superheaters
IP temperatures=279,461,565
2 IP economizers
2 IP superheaters
4 RH superheaters
LP temperatures=120,278,480
2 IP economizers
2 LP superheaters
```

name	mC	Pci	Pco	Tci	Tco	mH	Thi	Tho
RH2a	804264	454	452	929	1000	3600000	1125	1099
RH1b	804264	456	454	858	929	3600000	1099	1072
CRH	765683	460	460	725	717	3600000	0	0
RH1a	804264	458	456	788	858	3600000	1072	1045
DB	0	0	0	0	0	3614500	1045	1345
HPSH3a	795928	1530	1500	935	1000	3614500	1345	1319
LPSHa	66254	46	45	379	480	3614500	1319	1316
IPSHa	38581	470	460	513	565	3614500	1316	1315
HPSH1b	795928	1560	1530	871	935	3614500	1315	1287
HPSH2b	795928	1590	1560	806	871	3614500	1287	1257
RH2b	804264	460	458	717	788	3614500	1257	1231
HPSH2a	795928	1620	1590	741	806	3614500	1231	1197
HPSH1a	795928	1650	1620	676	741	3614500	1197	1156
HPSH3b	795928	1680	1650	612	676	3614500	1156	1093
HPEV	795928	1680	1680	607	612	3614500	1093	715
HPEC1b	795928	1830	1680	553	607	3614500	715	660
HPEC2b	795928	1980	1830	499	553	3614500	660	611
IPEV	38581	470	470	456	461	3614500	611	582
IPSHb	38581	480	470	461	513	3614500	582	581
HPEC2a	795928	2130	1980	445	499	3614500	581	534
IPECb	38581	500	480	368	456	3614500	534	531
HPEC1a	795928	2280	2130	391	445	3614500	531	486
LPSHb	66254	47	46	278	379	3614500	486	483
HPEC3a	795928	2430	2280	337	391	3614500	483	439
LPECa	66254	48	47	199	278	3614500	439	433
IPECa	38581	520	500	279	368	3614500	433	430
HPEC3b	795928	2580	2430	283	337	3614500	430	386
LPEV	66254	48	48	278	278	3614500	386	324
LPECb	900764	98	48	120	199	3614500	324	249

name	NTUc	NTUh	F	LMTD	UAc	UAh	UAf
RH2a	0.49	0.18	0.986	146.1	0.213	0.213	0.213
RH1b	0.37	0.14	0.992	190.7	0.162	0.162	0.162
RH1a	0.30	0.11	0.994	235.0	0.132	0.132	0.132
HPSH3a	0.18	0.07	0.998	364.1	0.089	0.089	0.089
LPSHa	0.11	0.01	1.000	886.8	0.004	0.012	0.004
IPSHa	0.07	0.01	1.000	776.2	0.002	0.012	0.002

```
HPSH1b 0.16 0.07 0.998 397.7 0.085 0.085 0.085
HPSH2b 0.15 0.07 0.998 433.5 0.084 0.084 0.084
RH2b    0.14 0.05 0.999 491.1 0.064 0.064 0.064
HPSH2a 0.15 0.08 0.998 440.2 0.092 0.092 0.092
HPSH1a 0.14 0.09 0.998 467.6 0.104 0.104 0.104
HPSH3b 0.14 0.13 0.997 480.6 0.153 0.153 0.153
HPEV              1.000 250.6             1.691
HPEC1b 0.52 0.54 0.960 107.8 0.575 0.575 0.575
HPEC2b 0.51 0.46 0.966 109.3 0.491 0.490 0.491
IPEV              1.000 137.9             0.215
IPSHb  0.57 0.02 0.998  92.4 0.017 0.017 0.017
HPEC2a 0.66 0.57 0.947  85.6 0.593 0.592 0.593
IPECb  0.76 0.03 0.996 115.8 0.032 0.032 0.032
HPEC1a 0.63 0.52 0.954  90.2 0.533 0.532 0.533
LPSHb  0.67 0.02 0.997 150.2 0.023 0.023 0.023
HPEC3a 0.58 0.48 0.960  96.2 0.481 0.481 0.481
LPECa  0.41 0.03 0.998 194.8 0.027 0.027 0.027
IPECa  0.87 0.03 0.995 102.3 0.035 0.035 0.035
HPEC3b 0.57 0.47 0.961  97.5 0.464 0.464 0.464
LPEV              1.000  71.9             0.853
LPECb  0.66 0.62 0.944 126.9 0.595 0.595 0.595
total HP/RH conductance=6.0
total IP conductance=0.3
total LP conductance=1.5
estimated net power 125955 kW
1001 converged, 1078 failed, 33 improved
```

Notice that the reheaters have been moved ahead of the HP superheaters. The IP and LP superheaters have also been moved up. Not surprisingly, the LP evaporator and economizer are still at the end. The differences are summarized in the following table:

Element Rearrangement			
Fms	Tdb	Tstk	kWnet
780,144	1316.3	239.0	124,637
786,893	1317.9	252.0	125,537
0.87%	1.6	13.0	0.72%

This rearrangement of elements yields a 0.87% increase in main steam flow, only a 1.6°F increase in duct burner exit temperature, a 13.0°F increase in stack temperature, and a 0.72% increase in estimated net power. Overall, it's not much. If you were hoping to double the main steam flow or pull the stack temperature down 100°F, that's not going to happen. Even with 33 steps of improvement over the initial configuration, only a very small increase in net power is achieved. This is at constant UA or heat exchange material cost. The heat release diagram for the final configuration can also be found in the spreadsheet arrange.xls and is shown on the next page.

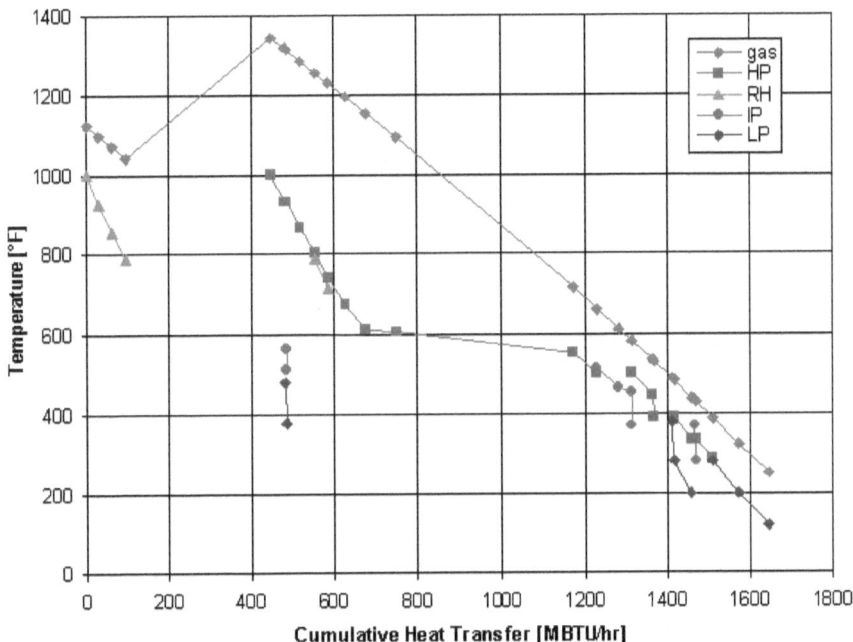

In spite of this being somewhat different from that of the initial configuration shown earlier in this chapter, the pinch point region between Q=1300 to 1500 MBTU/hr is very similar. As mentioned before, pinch points dominate the performance of a HRSG.

Chapter 8. Attemperation

Hotter is not necessarily better. As illustrated in the Mollier diagram and expansion line, you only need steam that is hot *enough*. Increasing the temperature beyond 1100°F/600°C can lead to material failures and doesn't provide a proportional increase in power output. More steam is always better, if it's hot enough and provided the equipment can handle it (i.e., adequate steam turbine swallowing capacity). I have tested more than one plant that had no reserve swallowing capacity, preventing them from making more money (selling more megawatt-hours) during times of peak demand, especially in cold months. This was an unwise design decision.

The worst case of under-performance I have ever seen is a system with far too much swallowing capacity—more than twice needed by the rest of the system. This resulted in low efficiency year round. This undesirable situation happened because a large international manufacturer (which will remain unnamed) winning the bid. This larger company had recently acquired a smaller company, wanted to integrate the two product lines, and utilize the existing stock. The company decided to simply pull one of the steam turbines off the shelf at this acquired company, not considering the consequences of this turbine being incorrectly sized for the application. This was a *business* decision, not an *engineering* one, so we know who got the short end of that stick! The turbine was actually quite well designed and fabricated, just the wrong size. As a result of this foolish decision, the company's reputation was damaged—in spite of having provided an excellent piece of equipment. If you need an alternator, a carburetor won't do the job, no matter how good it is.

Attemperation is accomplished by spraying feed water (liquid) into the superheated steam (vapor) to drop the temperature. This process naturally increases the mass flow rate. It is best to attemperate between the final two stages of superheat, which is one reason for having at least two HP and two RH superheaters in a HRSG. While it may be cheaper to construct and easier to maintain if attemperation is performed after the final superheat, as the steam leaves the HRSG, this isn't the best solution. You want this process to be sluggish and slowly responsive. It doesn't have to be a quick response. It may be more challenging to manage, but that's what control system logic is for. Adding feedwater before the final superheat assures complete mixing and also moderates temperature variations in the turbine. This adage is true: haste makes waste.

A note of caution... Should you get a bid package for a HRSG that indicates very large attemperation flows during normal operation, toss it in the trash. The components should be designed and fabricated to fit the application, which is driven by the gas turbine and steam turbine. You don't want to buy a HRSG that has been cobbled together with existing parts designed for some other system.

Chapter 9. Performance Testing

Let us begin by agreeing that it's not the HRSG manufacturer's fault if the GT isn't supplying sufficient flow and temperature. I've suffered through many arguments driven by a GT that provided too much flow and not enough temperature or vise versa. The HRSG design depends on both. Flow and temperature are not interchangeable. GT manufacturers love to talk about exhaust *energy*, meaning flow times temperature, but that's not how you design a HRSG. At the very least, there's an implied reference point for any and all enthalpies. Rate of energy supplied by the GT (actually thermal power) should at least be:

$$thermal\ power = \dot{m}\left(h_{EGT} - hR_{REF}\right) \tag{9.1}$$

but this still misses the point. Granted, gas turbines are volumetric flow devices, but HRSGs aren't—two completely different animals.

With this said, the steam flow produced by a HRSG is roughly proportional to the GT exhaust flow and the final superheat temperature is roughly proportional (though not in a multiplicative sense) to the exhaust temperature. But it's more complicated than this. Considering steam turbine requirements and attemperation, the final superheat temperature is somewhat fixed. In a performance test, you basically achieve it or you don't. The penalties per degree vary considerably, but these aren't particularly helpful and nobody comes away satisfied from a test failed. Net power output is approximately proportional to main steam flow so that penalties (or bonuses) based on flow rate are more meaningful than ones based on temperature. My advice is simple: Fix the GT and *then* retest the HRSG.

Code-Level Testing

Whether you are designing, fabricating, or purchasing a HRSG, you definitely want a defensible code-level performance test. The applicable code for testing HRSGs is ASME PTC-4.4, "Gas Turbine Heat Recovery Steam Generators," 2008. It is utterly pointless to perform a 4.4 test without also performing a test of the gas turbine to determine the exhaust flow. The applicable code for testing GT is ASME PTC-22, "Gas Turbines," 2014.

Some customers (end purchasers) are quite wary from being previously *burned* in an acquisition. This may motivate them to require all sorts of correction curves, adjustments, and test points. I can sympathize with this, much like buying a used car that looks nice, but falls apart as you're driving off the lot. Exorbitant penalties and endless requirements will not protect you from an experienced and dishonest car salesman. You can't possibly think of enough checks and tests. It is far wiser to contact previous purchasers of similar cars from the same dealer before laying out any cash. This also goes for buying a HRSG. Make sure the manufacturer knows what they're doing and has a proven track record.

There is a less common problem, which also must be considered and that is consistent information exchange. More than once I have seen a project that switched GTs (or STs) at some point and for various reasons (usually cost). This change was never communicated to the HRSG manufacturer (or the condenser or cooling tower manufacturer). The Engineering Procurement Contractor (EPC) dropped the proverbial ball and the HRSG, condenser, or cooling tower supplier faced penalties.

Accuracy is also important. I recall one case where the GT exhaust mole fractions of H_2O and CO_2 were reversed when transcribing information from one supplier to the other. I noticed this immediately, but that was after the plant had been built. Numerically, the H_2O and O_2 mole fractions are often closer than H_2O and CO_2, so that this mistake might be more likely overlooked. Water vapor has a much higher specific heat than any of the other constituents (see Appendix B). The end result of this discrepancy was insufficient energy entering the HRSG to produce the guaranteed steam, even if all were extracted such that the stack temperature were brought down to ambient. It was thermodynamically impossible to meet the guarantee. This was not the HRSG manufacturer's fault. The GT exhaust flow and temperature were considered in the guarantee, but not the composition. It wasn't the GT manufacturer's fault either. I know of two cases where the owner acquired GTs from the gray market and forced the change on the EPC. In one of those the HRSG manufacturer came out OK, but in the second one, they did not.

Single Pressure Combined Cycle Power Plant

You will find all of the file associated with a typical single pressure combined cycle power plant in the online archive in folder examples\CCPP named CCPP1.???. The two primary performance variables of interest are main steam flow and temperature. The other parameters are trivial (IP/LP steam flow and temperature. These are simply there to balance out the heat release diagram, provide non-condensable gas removal, and steady operation. They need to work properly, but the particular values don't really matter when considering performance. Reheat pressure drop is also important, but HRSGs rarely fail for this reason. If they ever do, you will have much bigger problems to worry with.

The main steam outlet temperature can vary and if it does, this will not be linear, as shown in the following figure:

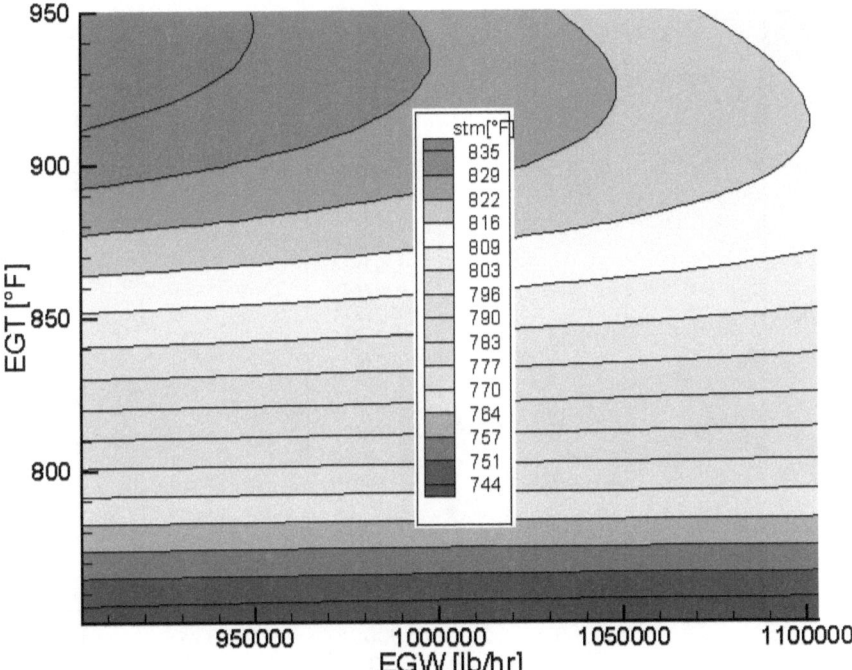

A simple linear correction curve will probably not do the job. This is the 21st century. Everything doesn't have to be a straight line. Excel® can handle a more complicated curve, as illustrated in the next chapter.

The main steam flow will be approximately bilinear (linear in two variables: z=a+bx+cy), as illustrated in this next figure:

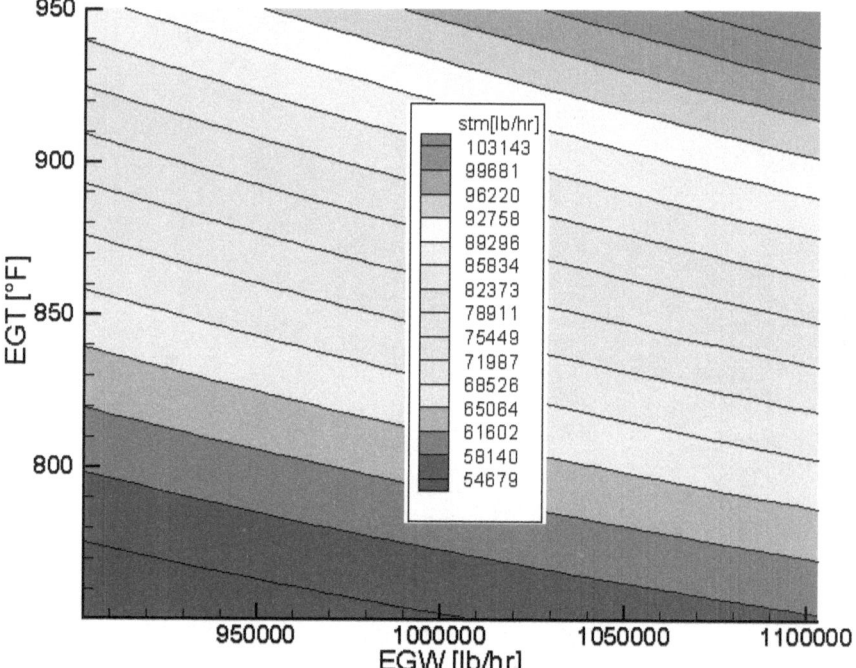

This is a simple relationship and easily corrected, that is, from the test point to the design point, as illustrated in the next chapter.

The net power output, which will be paying for the HRSG, is also approximately bilinear:

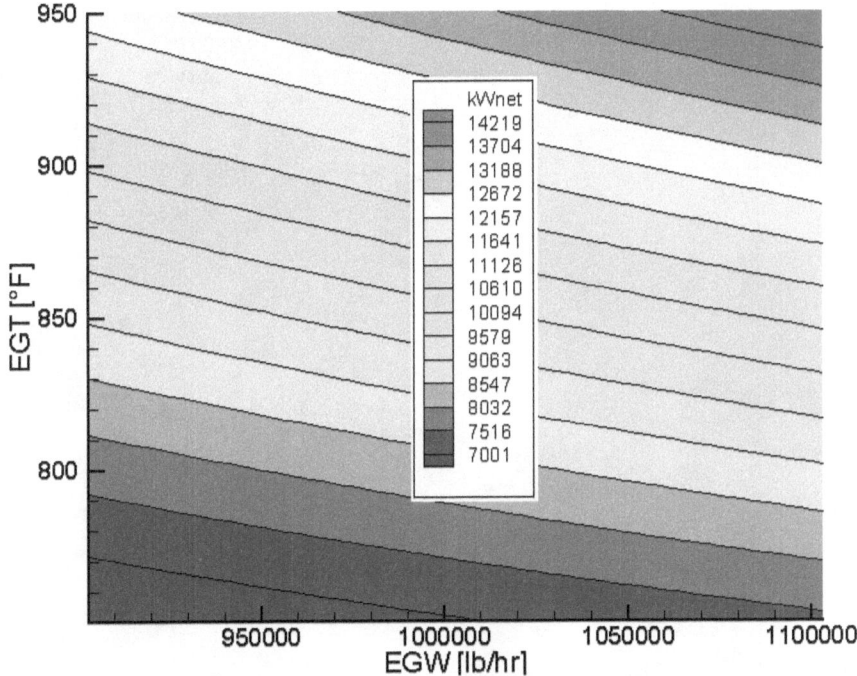

If you have GateCycle™ you can make adjustments to the model and see what happens. If you don't, you can work with the code (CCPP1.c) to try all sorts of things, including make it fit your design.

Three-Pressure Combined Cycle Power Plant

You will find all of the files associated with a typical three-pressure system (CCPP3.???) in the same folder. If you have GateCycle™ you can make runs or modify the code (CCPP3.c) to find the impact of various inputs and design parameters on output and performance measures.

Again, the variation in main steam temperature is nonlinear, though different from the previous system. This relationship will change with engine design., operation to control emissions, HRSG design, and ambient conditions (i.e., winter vs. summer).

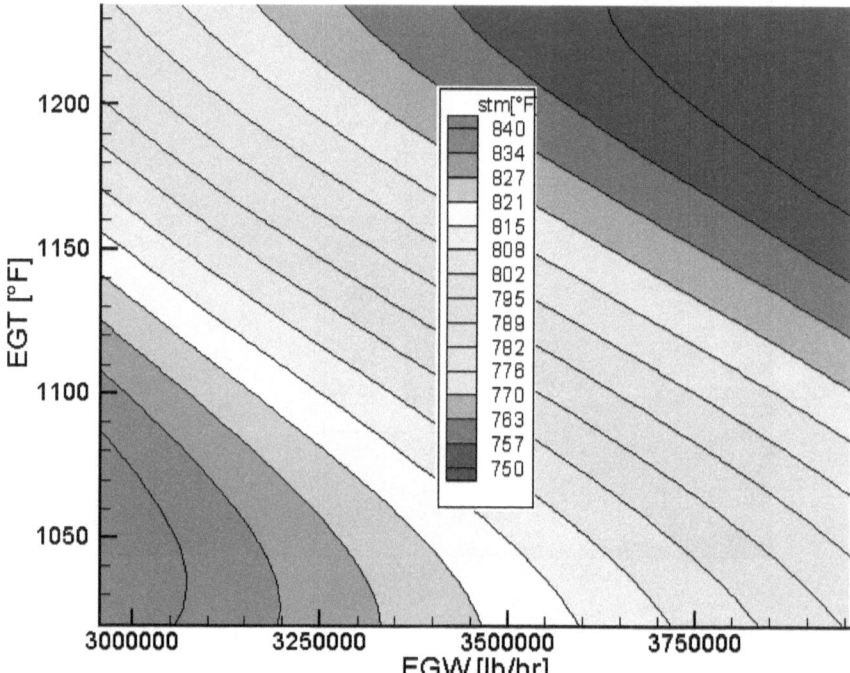

A linear correction will not suffice for this relationship either, but the steam flow and net power (of the steam tail) are again mostly bilinear.

The main steam flow in this case is almost exactly bilinear:

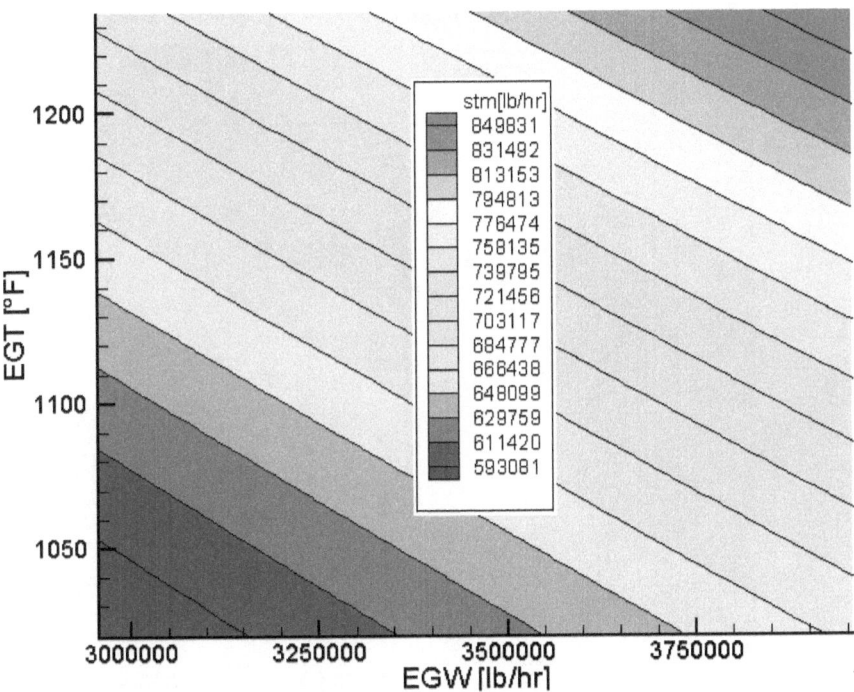

A bilinear correction or two linear ones would be adequate in this case.

The net power output of the steam tail is not quite bilinear. I would recommend using a second order correction ($z=a+b*x+c*y+d*x*y+e*x^2+f*y^2$):

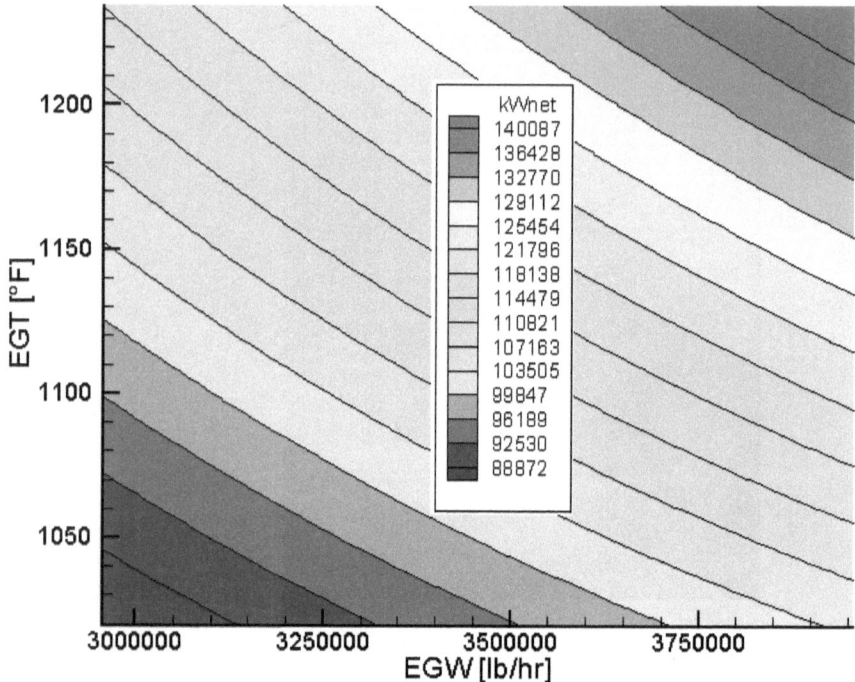

Chapter 10. GT and HRSG Heat Balances

As stated previously, you will want to perform a GT (PTC-22) and a HRSG (PTC-4.4) test. While the HRSG test does provide an estimate of the GT exhaust flow, there are far more assumptions and things taken for granted that are not necessarily the case. If the two don't agree, this provides awareness and a path forward to resolve the differences. A basic three-pressure system is illustrated below:

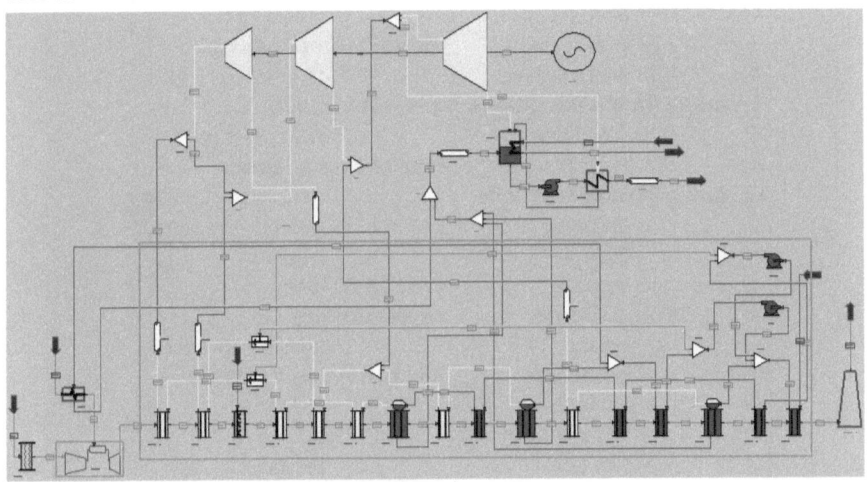

There are two boundaries (control volumes): 1) around the GT and 2) around the HRSG. We will first consider the GT. The most important thing to remember here is that the ASHRAE (moist air) properties are NOT consistent with the NASA (exhaust gas) OR the ASME/IAPWS (steam) properties. These have three different reference points and you can't just subtract the enthalpies without correcting for this. The same goes for the heats of formation (and consequently heats of combustion). These have a specific reference and you must not ignore this.

> *If your results (GT exhaust flow/HRSG heat loss) change in the slightest when you change the reference temperature, then your calculations are simply wrong! The reference temperature is an arbitrary value and the real world doesn't know or care what your reference temperature is.*

You will find all of the necessary calculations in the online archive in folder examples\testing in the spreadsheet GT_heat_balance.xls. User inputs are blue and calculations are violet.

GT Heat Balance

	A	B	C	D
1				**GAS TURBINE HE**
2	**INPUTS**		**INPUTS Contd.**	
3	**reference temperature**		**injection flow & conditions**	
4	77	ref. temp.["F]	10,000	flow [lbm/hr]
5	**ambient conditions**		150	pres.[psia]
6	14.540	amb. pres.[psia]	450	temp.["F]
7	95	amb. temp.["F]	**exhaust temperature**	
8	43%	amb. rel. hum.	1108	temp.["F]
9	**bleed flow & conditions**		**generator output**	
10	5,000	flow [lbm/hr]	167,700	net power [kW]
11	600	temp.["F]	**efficiencies & losses**	
12	**fuel gas mole fractions**		98.89%	generator efficiency
13	95.19%	Methane	100.00%	gearbox efficiency
14	2.48%	Ethane	0.82%	heat loss
15	0.44%	Propane	**CALCULATIONS**	
16	0.09%	IsoButane	**psychrometry**	
17	0.09%	Butane	0.015457	humidity [lb-wetr/lb-dry]
18	0.04%	IsoPentane	**fuel gas mass fractions**	
19	0.02%	Pentane	89.66%	Methane
20	0.06%	Hexane	4.38%	Ethane
21	0.51%	Nitrogen	1.13%	Propane
22	0.00%	Carbon Monoxide	0.32%	IsoButane
23	1.07%	Carbon Dioxide	0.32%	Butane
24	0.00%	Water	0.16%	IsoPentane
25	0.00%	Hydrogen Sulfide	0.10%	Pentane
26	0.00%	Hydrogen	0.31%	Hexane
27	0.00%	Helium	0.84%	Nitrogen
28	0.00%	Oxygen	0.00%	Carbon Monoxide
29	0.00%	Argon	2.77%	Carbon Dioxide
30	100.00%	total	0.00%	Water
31	17.031	mol. wt.	0.00%	Hydrogen Sulfide
32	**fuel gas flow & conditions**		0.00%	Hydrogen
33	76,149	flow [lbm/hr]	0.00%	Helium
34	440	pres.[psia]	0.00%	Oxygen
35	365	temp.["F]	0.00%	Argon
36			100.00%	total

You will need to measure the barometric pressure, ambient temperature, and ambient humidity. Note that humidity sensors are not particularly accurate, which is why wet-bulb RTDs are most often used. A dew-point device is also quite accurate, but they don't work well dangling from a rope and swinging in the breeze.

You can enter zero for the compressor bleed if there is none. If there is water or steam injection, you must accurately measure the flow, pressure, and temperature. Measuring the exhaust temperature requires a rack of thermocouples or high temperature RTDs. As these are in a high velocity flow zone, special care is required affixing them and protecting the cables. Thermocouples are far more durable.

A precision power meter is required to determine the generator output and power factor. This is connected to special voltage and current transformers. The generator efficiency is obtained from curves, which are usually adequate for this purpose. The gearbox efficiency and heat loss cannot be measured in such a test, so you will need these values from the manufacturer, which they will have measured on a test stand.

Fuel gas flow is measured with a sharp-faced orifice plate or Coriolis meter. Do not accept anything else for fuel flow measurement, specifically, DO NOT accept an annubar or any other type of out-of-a-box flow meter, no matter how good someone tells you it is or how much it costs. While these devices may be accurate under all the right conditions, you don't know that to be the case! If you're not going to accurately measure the fuel flow, then don't bother performing a test because clearly you don't care (see Appendix F).

Fuel composition is measured with a calibrated gas chromatograph. You will need calibration gases and a unit that is much too sensitive to hang in a rack at a power plant between a fire hose and an axe. Online chromatographs (i.e., continuously operating) are OK for gross characterization and day-to-day monitoring of plant performance. These may even be quite accurate when they're first installed after calibration. They are not suitable for a performance test.

Fuel composition is entered by mole ratio, as this is customary. Mass fractions are needed for the calculations. The mass fractions are not inputs; rather, these are calculated from the mole fractions and molecular weights.

Exhaust gas mole fractions are calculated from the ambient plus fuel minus bleed plus water or steam injection. Both mass and mole fractions, as are mass and molar flow rates.

Steam enthalpies in the spreadsheet are calculated using ASME 1967 properties, which are supplied as macros. You could change these to something else, for instance, using the AllSteam Excel® Add-In. The gas enthalpies are calculated using NASA Glenn and also PTC-22, which should be consistent. These calculations are also implemented as macros. There are also 3800 entries on the NASA tab.

E	F	G	H

ΛT BALANCE CALCULATIONS

CALCULATIONS Contd.		CALCULATIONS Contd.	
enthalpy of dry air @Tin		**reactants [lbm/hr]**	
4.32	NASA enth.[BTU/lbm]	204,579	Carbon Dioxide
4.32	refd. enth.[BTU/lbm]	162,158	Water
enthalpy of dry air @Tbleed		640	Nitrogen
127.66	NASA enth.[BTU/lbm]	0	Sulfur Dioxide
127.66	refd. enth.[BTU/lbm]	0	Argon
enthalpy of dry air @Texh		0	Helium
258.93	NASA enth.[BTU/lbm]	291,227	stoichiometric O2
258.93	refd. enth.[BTU/lbm]	1,258,445	dry air [lbm/hr]
enthalpy of moisture @Tin		1,277,898	moist air [lbm/hr]
8.02	NASA enth.[BTU/lbm]	**products [lbm/hr]**	
8.02	refd. enth.[BTU/lbm]	205,183	Carbon Dioxide
enthalpy of moisture @Tbleed		162,158	Water
240.83	NASA enth.[BTU/lbm]	951,008	Nitrogen
240.83	refd. enth.[BTU/lbm]	0	Sulfur Dioxide
enthalpy of moisture @Texh		16,245	Argon
495.94	NASA enth.[BTU/lbm]	1	Helium
495.94	refd. enth.[BTU/lbm]	1,334,595	total [lbm/hr]
injection enthalpies		**enthalpy of products @Texh**	
vapor	state	290.72	NASA enth.[BTU/lbm]
1247.36	ASME enth.[BTU/lbm]	290.72	refd. enth.[BTU/lbm]
151.89	NASA enth.[BTU/lbm]	**air calculations**	
151.89	refd. enth.[BTU/lbm]	2,277,614	dry excess air (NASA) [lbm/hr]
fuel gas enthalpies & LHV		2,277,614	dry excess air (refd.) [lbm/hr]
164.00	NASA enth.[BTU/lbm]	0	error (refd.-NASA) [lbm/hr]
164.00	refd. enth.[BTU/lbm]	2,312,819	moist excess air [lbm/hr]
20,646	LHV [BTU/lbm]	3,595,717	moist total in air [lbm/hr]
0	LHV adj. [BTU/lbm]	3,676,866	moist total exh. air [lbm/hr]
20,646	adj. LHV [BTU/lbm]	**enthalpy of exhaust @Texh**	
power & heat loss		274.64	NASA enth.[BTU/lbm]
578.67	Shaft [MBTU/hr]	274.64	refd. enth.[BTU/lbm]
12.82	Heat Loss [MBTU/hr]		

This is a non-iterative calculation, that is, there is no button to push or residual to minimize. The entire system of algebraic equations are solved simultaneously to obtain the result, in this case, dry air flow. If you happen to have Maple™ or can read a Maple™ worksheet, the equations are in the same folder in file GT_heat_balance.mws. The final equation is Equation 10.1:

$$m_{excess\_dry} = \cfrac{\begin{pmatrix} m_{fuel}\left(LHV + h_{fuel,\,Tfuel} - h_{products,\,Texh}\right) - Q_{loss} - W_{shaft} \\[2mm] -\dfrac{\left(h_{air,\,Tbleed} - h_{air,\,Tin} + \omega\left(h_{H2O,\,Tbleed} - h_{H2O,\,Tin}\right)\right)m_{bleed\_total}}{1+\omega} \\[2mm] - m_{comb\_dry}\left(h_{products,\,Texh} - h_{air,\,Tin} + \omega\left(h_{H2O,\,Texh} - h_{H2O,\,Tin}\right)\right) \\[2mm] - m_{inject}\left(h_{H2O,\,Texh} - h_{H2O,\,Tinj}\right) \end{pmatrix}}{\left(h_{air,\,Texh} - h_{air,\,Tin} + \omega\left(h_{H2O,\,Texh} - h_{H2O,\,Tin}\right)\right)}$$

The calculations continue across the sheet. If you change any of the blue values, the results will quickly update. Excel® handles the sequential computations and formulas automatically.

EXHAUST (after injection)		Add-In Macro Test	
mass flow [lbm/hr]		3,676,892	exhaust flow [lbm/hr]
2,671,045	Nitrogen	73.2115%	Nitrogen
206,277	Carbon Dioxide	3.5988%	Carbon Dioxide
226,816	Water	9.6640%	Water
527,081	Oxygen	12.6479%	Oxygen
0	Sulfur Dioxide	0.0000%	Sulfur Dioxide
45,645	Argon	0.8773%	Argon
3	Helium	0.0005%	Helium
3,676,867	total [lbm/hr]	**mole fractions**	
mass fractions		73.2093%	Nitrogen
72.6446%	Nitrogen	3.5988%	Carbon Dioxide
5.6101%	Carbon Dioxide	9.6668%	Water
6.1687%	Water	12.6472%	Oxygen
14.3351%	Oxygen	0.0000%	Sulfur Dioxide
0.0000%	Sulfur Dioxide	0.8773%	Argon
1.2414%	Argon	0.0005%	Helium
0.0001%	Helium	100.0000%	total
100.0000%	total		
molar flow [moles/hr]			
95,349	Nitrogen		
4,687	Carbon Dioxide		
12,590	Water		
16,472	Oxygen		
0	Sulfur Dioxide		
1,143	Argon		
1	Helium		
130,241	total [mole/hr]		
28.231	mol. wt.		

The fuel properties are on the Fuels tab of the spreadsheet. The elements and required oxygen are shown, along with the lower heating value. Note that the higher heating value is a meaningless term. Higher heating is based on water vapor in the combustion products existing in the liquid state, which isn't going to happen unless you're operating on Antarctica in a blizzard. People who insist on using higher heating value clearly don't understand thermodynamics.

	A	B	C	D	E	F	G	H	I	J	K
1											
2	**Properties of Fuels (from NIST)**										
3											
4	**Name**	**Formula**	**LHV**	**C**	**H**	**N**	**O**	**S**	**Ar**	**He**	**O2**
5	Methane	CH_4	21,512	1	4	0	0	0	0	0	-2.0
6	Ethane	C_2H_6	20,429	2	6	0	0	0	0	0	-3.5
7	Propane	C_3H_8	19,922	3	8	0	0	0	0	0	-5.0
8	IsoButane	C_4H_{10}	19,590	4	10	0	0	0	0	0	-6.5
9	Butane	C_4H_{10}	19,658	4	10	0	0	0	0	0	-6.5
10	IsoPentane	C_5H_{12}	19,456	5	12	0	0	0	0	0	-8.0
11	Pentane	C_5H_{12}	19,497	5	12	0	0	0	0	0	-8.0
12	Hexane	C_6H_{14}	19,393	6	14	0	0	0	0	0	-9.5
13	Nitrogen	N_2	0	0	0	2	0	0	0	0	0.0
14	Carbon Monoxide	CO	4,342	1	0	0	1	0	0	0	-0.5
15	Carbon Dioxide	CO_2	0	1	0	0	2	0	0	0	0.0
16	Water	H_2O	0	0	2	0	1	0	0	0	0.0
17	Hydrogen Sulfide	H_2S	6,534	0	2	0	0	1	0	0	-1.5
18	Hydrogen	H_2	51,567	0	2	0	0	0	0	0	-0.5
19	Helium	He	0	0	0	0	0	0	0	1	0.0
20	Oxygen	O_2	0	0	0	0	2	0	0	0	1.0
21	Argon	Ar	0	0	0	0	0	0	1	0	0.0
22	per mole			1	2	2	1	1	1	1	1

Heating values and heats of formation are intimately linked. They also have a reference point (pressure and temperature). These are not independent. Do not change them unless you know what you're doing. The spreadsheet will perform the calculations for you. It is interesting to note that the heating values change when you change the reference condition on the first tab. If they don't, then you won't get the same answer and your calculations will be incorrect. NASA had to burn and/or ionize and split these substances in order to build the tables. You don't read heating value from a gas chromatograph.

												net	NIST
							based on NASA table						
reactants					products							NASA	NASA
by mole		by mass			by mole			by mass				NASA	NASA
O2	Fuel	O2	Fuel	Hf	H2O	CO2	SO2	H2O	CO2	SO2	Hf	LHV	diff.
2.0	1.0	79.96%	20.04%	-400.7	2.0	1.0	0.0	45.02%	54.98%	0.00%	-4711.5	21,508	0.02%
3.5	1.0	78.83%	21.17%	-253.8	3.0	2.0	0.0	38.04%	61.96%	0.00%	-4577.2	20,427	0.01%
5.0	1.0	78.39%	21.61%	-220.5	4.0	3.0	0.0	35.31%	64.69%	0.00%	-4524.5	19,920	0.01%
6.5	1.0	78.16%	21.84%	-218.1	5.0	4.0	0.0	33.85%	66.15%	0.00%	-4496.4	19,588	0.01%
6.5	1.0	78.16%	21.84%	-203.2	5.0	4.0	0.0	33.85%	66.15%	0.00%	-4496.4	19,656	0.01%
8.0	1.0	78.01%	21.99%	-201.4	6.0	5.0	0.0	32.94%	67.06%	0.00%	-4478.9	19,454	0.01%
8.0	1.0	78.01%	21.99%	-192.3	6.0	5.0	0.0	32.94%	67.06%	0.00%	-4478.9	19,496	0.01%
9.5	1.0	77.91%	22.09%	-183.9	7.0	6.0	0.0	32.32%	67.68%	0.00%	-4466.9	19,392	0.01%
0.5	1.0	36.35%	63.65%	-1079.8	0.0	1.0	0.0	0.00%	100.00%	0.00%	-3844.1	4,343	-0.03%
1.5	1.0	58.48%	41.52%	-107.9	1.0	0.0	1.0	21.95%	0.00%	78.05%	-2821.3	6,535	-0.02%
0.5	1.0	88.81%	11.19%	0.0	1.0	0.0	0.0	100.00%	0.00%	0.00%	-5771.0	51,574	-0.01%

While it is generally assumed that the composition of air is uniform all over the Earth and at all times, this is not exactly true. It is a fair assumption though and probably better than some of the others, including: gear and heat loss as well as generator efficiency. Air composition is on a separate tab and linked to the calculations. Note that the total must be 100%. I have seen many discrepancies on the Web where various constituencies don't sum to unity. In this spreadsheet, the trace amounts of noble gases are lumped together. These don't react and the specific heats are flat, so it's not a significant consideration.

	A	B	C	D
1	**Constituents of Air from the CRC Handbook**			
2	**Chemistry & Physics, 1997 Edition.**			
3	**air constituents by mole**			
4	78.08400%	Nitrogen		
5	20.94760%	Oxygen		
6	0.93400%	Argon		
7	0.03140%	Carbon Dioxide		
8	0.00182%	Neon		
9	0.00020%	Methane		
10	0.00052%	Helium		
11	0.00011%	Krypton		
12	0.00005%	Hydrogen		
13	0.00001%	Xenon		
14	99.99971%	total		
15	**consolidated & normalized by mole**			
16	78.08422%	Nitrogen		
17	20.94766%	Oxygen		
18	0.93594%	Argon, Neon, Krypton & Xenon		
19	0.03160%	Carbon Dioxide & Methane		
20	0.00057%	Helium & Hydrogen		
21	100.00000%	total		
22	**dry air constituents by mole**			
23	78.0842%	Nitrogen		
24	20.9477%	Oxygen		
25	0.9359%	Argon		
26	0.0316%	Carbon Dioxide		
27	0.0006%	Helium		
28	100.0000%	total		
29	28.965	mol. wt.		
30	**dry air constituents by mass**			
31	75.5192%	Nitrogen		
32	23.1418%	Oxygen		
33	1.2908%	Argon		
34	0.0480%	Carbon Dioxide		
35	0.0001%	Helium		
36	100.0000%	total		

HRSG Heat Balance

The HRSG heat balance is in the same spreadsheet on a different tab. Again, the user inputs are blue and calculations violet. An energy balance is needed to calculate the heat loss from the HRSG. Don't assume this value, measure everything else and then calculate it. Even if you took infrared pictures all over the HRSG and stitched them together into a 3D shell, you still couldn't accurately calculate the heat loss, because there's no way to accurately estimate the convective and radiative heat transfer coefficients.

	A	B	C	D
1				**HRSG HEAT BALA**
2	**INPUTS**		**INPUTS Contd.**	
3	**operating pressures**		**pump efficiencies**	
4	245	LP FW pres.[psia]	75%	recirc. pump eff.
5	0	Recirc. dP [psi]	70%	IP pump eff.
6	81	LP Drum pres.[psia]	65%	HP pump eff.
7	78	LP Steam pres.[psia]	**CALCULATIONS**	
8	615	IP FW pres.[psia]	**flows**	
9	564	IP Drum pres.[psia]	781,530	total FW flow [lbm/hr]
10	557	CRH pres.[psia]	781,530	FW+Recirc. [lbm/hr]
11	527	HRH pres.[psia]	99,580	IP FW [lbm/hr]
12	2240	HP FW pres.[psia]	682,570	HRH [lbm/hr]
13	2098	HP Drum pres.[psia]	644,470	HP FW [lbm/hr]
14	1942	HP Steam pres.[psia]	0	total blowdown [lbm/hr]
15	**operating temperatures**		1,405,970	steam entering [lbm/hr]
16	182	Stack temp.[°F]	1,405,970	steam leaving [lbm/hr]
17	111	LP FW Inlet temp.[°F]	3,689,966	stack [lbm/hr]
18	633	LP Steam temp.[°F]	**temperatures**	
19	727	CRH temp.[°F]	313	LP Drum [°F]
20	1057	HRH temp.[°F]	314	IP FW [°F]
21	1056	Main Steam temp.[°F]	480	IP Drum [°F]
22	60	DB Fuel temp.[°F]	319	HP FW [°F]
23	**operating flows**		643	HP Drum [°F]
24	0	recirc. flow [lbm/hr]	**steam enthalpies**	
25	0	LP Blowdown [lbm/hr]	79.6	LP FW [BTU/lbm]
26	37,480	LP Steam [lbm/hr]	283.0	LP Blowdown [BTU/lbm]
27	0	IP Blowdown [lbm/hr]	1347.4	LP Steam [BTU/lbm]
28	41,450	IP FW to FGHX [lbm/hr]	285.2	IP FW [BTU/lbm]
29	58,130	IP Steam [lbm/hr]	464.0	IP Blowdown [BTU/lbm]
30	624,440	CRH flow [lbm/hr]	1369.7	CRH Steam [BTU/lbm]
31	0	HP Blowdown [lbm/hr]	1550.4	HRH Steam [BTU/lbm]
32	644,470	HP Steam [lbm/hr]	292.9	HP FW [BTU/lbm]
33	13,100	DB Fuel flow [lbm/hr]	683.6	HP Blowdown [BTU/lbm]
34			1510.8	HP Steam [BTU/lbm]

Feedwater flow, as measured by a calibrated nozzle of ASME grade, is far more accurate than anything you could possibly do with the steam. Always measure the liquid flows and calculate the vapor flows. See Appendix F for more details.

E	F	G	H

NCE CALCULATIONS

CALCULATIONS Contd.		CALCULATIONS Contd.	
fuel gas enthalpies & LHV		**stack mass fractions**	
-8.65	NASA enth.[BTU/lbm]	72.3897%	Nitrogen
-8.65	refd. enth.[BTU/lbm]	6.5440%	Carbon Dioxide
20,646	LHV [BTU/lbm]	6.9028%	Water
0	LHV adj. [BTU/lbm]	12.9264%	Oxygen
20,646	adj. LHV [BTU/lbm]	0.0000%	Sulfur Dioxide
DB fuel [mole/hr]		1.2370%	Argon
800	Carbon Dioxide	0.0001%	Helium
1,548	Water	100.0000%	total
4	Nitrogen	**stack enthalpies**	
-1,566	Oxygen	26.72	NASA enth.[BTU/lbm]
0	Sulfur Dioxide	26.72	refd. enth.[BTU/lbm]
0	Argon	**heat transfer to steam**	
0	Helium	47.52	to LP [MBTU/hr]
stack [mole/hr]		214.27	to IP+RH [MBTU/hr]
95,353	Nitrogen	922.37	to HP [MBTU/hr]
5,487	Carbon Dioxide	**heat input from pumps**	
14,139	Water	0.00	recirc. [MBTU/hr]
14,906	Oxygen	0.23	IP [MBTU/hr]
0	Sulfur Dioxide	6.35	HP [MBTU/hr]
1,143	Argon	**ENERGY BALANCE**	
1	Helium	1009.81	GT exh. [MBTU/hr]
131,028	total [mole/hr]	-0.11	Fuel (sensible) [MBTU/hr]
28.162	mol. wt.	270.47	Fuel (LHV) [MBTU/hr]
stack mole fractions		6.57	pumps [MBTU/hr]
72.7729%	Nitrogen	1286.74	total in [MBTU/hr]
4.1875%	Carbon Dioxide	1184.16	steam [MBTU/hr]
10.7906%	Water	98.59	stack [MBTU/hr]
11.3764%	Oxygen	1282.74	total out [MBTU/hr]
0.0000%	Sulfur Dioxide	3.99	loss [MBTU/hr]
0.8720%	Argon	0.34%	loss
0.0005%	Helium		
100.0000%	total		

Chapter 11. Correcting Test Results

We have already seen the most important corrections in Chapter 9. Some contracts call for feed water temperature and even ambient corrections, but these are rarely meaningful. To investigate this we return to the single-pressure model (CCPP1). We can use GateCycle™ to vary the condenser cooling water temperature, which will result in a change in condenser backpressure and condensate temperature that will be felt at the LP economizer. Consider the following table from CCPP1.xls:

Tccw	Tfw	Fms	Tms	power
°F	°F	lb/hr	°F	kW
40	70.2	73,256	805.165	10,111
50	78.4	73,256	805.165	10,058
60	87.2	73,256	805.165	9,932
70	96.3	73,256	805.165	9,733
80	106.0	73,256	805.165	9,476
90	116.0	73,256	805.165	9,178
100	125.9	73,256	805.165	8,862
110	136.0	73,256	805.165	8,521
120	146.1	73,256	805.165	8,175

Even though the CCW temperature varies from 40°F to 120° and the feed water inlet temperature rises accordingly from 70.2°F to 146.1°F, there is no change in main steam flow (73,256 lb/hr) or main steam temperature (805.165°F). The ST generator output does vary, but this is not a result of changes in HRSG operation and so there should be no correction. That is, we cannot correct for feed water temperature based on this model. The same thing happens with the three-pressure model (see CCPP3.xls):

Tccw	Tfw	Fms	Tms	power
°F	°F	lb/hr	°F	kW
40	104.7	690,191	789.2	115,447
50	107.9	690,191	789.2	115,367
60	112.3	690,206	789.2	115,131
70	117.8	690,191	789.2	114,520
80	124.9	690,191	789.2	113,226
90	132.8	690,191	789.2	110,903
100	140.9	690,190	789.2	107,705
110	149.3	690,203	789.2	104,055
120	157.8	690,192	789.2	100,109

This problematic result (no impact from some parameter that should result in at least a small change) is one of the flaws in an otherwise excellent tool. The whole convergence and iterative calculation process within the program is disappointing. The GateCycle™ program was written so that it obtains final results very quickly but at the cost of accuracy. While there are adjustable convergence criteria, changing these is pointless. They're already set to the limits and it simply won't do any better. There is also the opposite problem of yielding differences when there should be none. For instance, if you run ambient temperatures of 50, 60, 70, 80, 90 you will get slightly different results than for 90, 80, 70, 60, and 50. Such should never occur with any software, unless it's a Monte Carlo simulation using random numbers.

The impact of feed water temperature on main steam flow is not zero, but it is quite small. I have been required on more than one occasion to provide a non-zero correction. While this may be challenging with GateCycle™, it is possible and may require composing special macros and other advanced techniques. The figure below is for an actual plant, which I tested and will convey the magnitude of the appropriate correction.

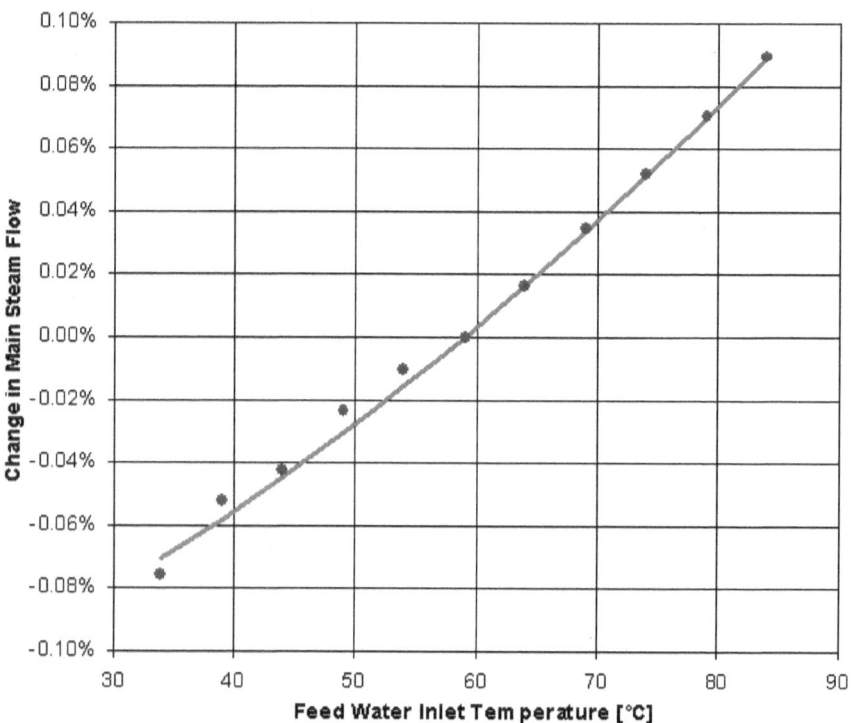

Single-Pressure HRSG Test

Typical summary test results are listed in the following table. The GT exhaust temperature and flow are determined by the PTC-22 test on a separate tab and brought in with linked formulas. The main steam flow and temperature are determined by the PTC-4.4 test and also brought in with links. The main steam flow is calculated from the measured feedwater flow and not a direct measurement of the steam, as this would be much less accurate. It is customary to average four one-hour test intervals.

SINGLE PRESSURE HRSG TEST							
INPUTS	units	Design	Test1	Test2	Test3	Test4	Avg.
GT exhaust flow	lb/hr	1,003,069	1,013,099	1,008,084	1,018,115	1,020,622	1,012,598
GT exhaust temp.	°F	850.2	848.2	852.6	847.4	848.9	849.5
main steam flow	lb/hr	73,275	74,008	73,641	74,374	74,557	73,971
main steam temp.	°F	805.1	834.4	809.0	777.2	834.2	812.0
CORRECTIONS	units	Design	Test1	Test2	Test3	Test4	Avg.
main steam flow	lb/hr	0	-191	-836	-324	-805	-431
main steam temp.	°F	0.0	1.4	-0.8	2.1	1.5	0.8
CORRECTED	units	Design	Test1	Test2	Test3	Test4	Avg.
main steam flow	lb/hr	73,275	73,817	72,805	74,050	73,752	73,540
main steam temp.	°F	805.1	835.8	808.2	779.3	835.7	812.8
TEST RESULT	units	Design	Test1	Test2	Test3	Test4	Avg.
main steam flow	-	N/A	PASS	FAIL	PASS	PASS	PASS
main steam temp.	-	N/A	PASS	PASS	FAIL	PASS	PASS

Note that the corrections are opposite performance calculations. Performance calculations answer the question: what is the expected flow or temperature under some particular conditions? Test corrections answer the question: what would the expected flow or temperature be at the guarantee conditions? If the as-tested GT exhaust flow is greater than expected, the measured steam flow is adjusted downward—likewise for the as-tested GT exhaust temperature. This process assures that one supplier is not penalized (or credited) for the product of another equipment supplier.

Overall Plant vs. Component Testing

As Dr. Keith Kirkpatrick[8] likes to say, "A 4 plus a 6 test does not equal a 46 test!" ASME PTC-4 and PTC-6 are the standard tests for conventional boilers and steam turbines, respectively. PTC-46 is a test for overall plant performance. Do not think that combining the first two will be an adequate replacement for the third. The same goes for a PTC-22 (gas turbine) and PTC-4.4 (HRSG) test. Each test is an evaluation of the equipment provided by a different supplier. The Engineering Procurement Contractor (EPC) should be responsible for everything, including the individual components (GT, HRSG, condenser, cooling tower, fire protection system, and everything else). Do not purchase a power plant thinking that all will be well and if it isn't, you'll just sue each and

---

[8] VP and Director of Testing for McHale Performance https://www.mchale.com/

every supplier. You will have to operate it for years to come, long after the lawyers have retired to the beach.

In recent years, I have noticed a disturbing trend... Several gas turbine manufacturers have taken on the role of EPC, figuring that a lot of money has been flowing into other pockets, which should have been theirs. I have yet to see this turn out well for the owner. What happens is common in human experience and not unique to power plant construction. One person who knows nothing of what another does, considers themselves oh so very much smarter and more hard-working, thinking they could do the other's job in their sleep. This is merely ignorance. It is shocking to see how much falls through the cracks when someone who has never done the work of an EPC takes on the roll.

This is a true story...
When we showed up to test one plant, everything was still in the crates. There wasn't so much as a parking lot, fence, guard shack, power line, sewer line, or porta potty! The inexperienced EPC thought the gas turbine supplier (a very reputable and experienced manufacturer) would set up their own equipment and not just ship it to the site and drop it off the back of a truck.

Three-Pressure HRSG Test

The following table shows typical test data and corrections for the three-pressure example (CCPP3.*):

THREE-PRESSURE HRSG TEST							
INPUTS	units	Design	Test1	Test2	Test3	Test4	Avg.
GT exhaust flow	lb/hr	3,456,937	3,455,844	3,459,590	3,454,478	3,456,881	3,456,746
GT exhaust temp.	°F	1127.0	1129.1	1127.6	1129.5	1127.6	1128.2
main steam flow	lb/hr	697,679	705,436	704,378	696,971	700,897	701,072
main steam temp.	°F	791.5	796.7	785.4	796.2	797.5	793.5
CORRECTIONS	units	Design	Test1	Test2	Test3	Test4	Avg.
main steam flow	lb/hr	0	-1,576	-798	-1,740	-479	-919
main steam temp.	°F	0.0	0.8	0.4	0.9	0.2	0.5
CORRECTED	units	Design	Test1	Test2	Test3	Test4	Avg.
main steam flow	lb/hr	697,679	703,860	703,580	695,231	700,418	700,153
main steam temp.	°F	791.5	797.5	785.8	797.1	797.7	793.9
TEST RESULT	units	Design	Test1	Test2	Test3	Test4	Avg.
main steam flow	-	N/A	PASS	PASS	FAIL	PASS	PASS
main steam temp.	-	N/A	PASS	FAIL	PASS	PASS	PASS

It is not common to fail one out of four test periods and pass the other three. This is only shown here for example. Typically, a HRSG either passes of fails miserably, without much in between. Most manufacturers build in a small margin, although some EPCs will presume this to be the case and may incorporate it or add a little of their own.

Appendix A: Steam Properties

The thermodynamic and transport properties of steam are defined in the various reports of the International Association for the Properties of Water and Steam. This is a link to their web site http://www.iapws.org/

Three primary versions are relevant in this case: IF-67[9], SF-95[10], and IF-97[11]. The designations IF and SF indicate Industrial and Scientific Formulations, respectively. The industrial formulations are in terms of pressure and temperature, while the scientific formulation is in terms of temperature and density. While the industrial formulations are less accurate and mathematically sloppy, they have been embraced throughout the industry. The notion that the scientific formulation is burdensome and unnecessarily complicated is, of course, made irrelevant by modern computers and well-crafted software. The IF-67 is used by some manufacturers (most notably G.E.) because it has been built into the basis of some archaic software and empirical correlations. There is precious little difference between the '67 and '97 industrial formulations when it comes to HRSG performance. The choice is most often driven by the steam turbine manufacturer and not by the HRSG manufacturer.

Excel® Add-In for Academic (Non-Commercial) Use Only

The most extensive text on Excel® Add-Ins is that of Steve Dalton.[12] This excellent document can be found in PDF form at several locations on the Web. Add-Ins (and, for that matter software) development is beyond the scope of this text. The source code provided can be compiled for either bitness. In fact, the batch file provided produces both. I have provided these steam properties and associated Excel® Add-In for academic (i.e., non-commercial) use only. While I have contributed considerable effort to translate these property functions into C and build them into the Add-In wrapper, the underlying code (mostly in FORTRAN) is the work of others (as noted) and can be found on the Web. There are several agents selling similar code for commercial. I do not want to compete with these. I also assume no liability for the use of this software. See the folder examples\AllSteam. You will also find a spreadsheet that generates a Mollier diagram in either English or SI units for any of the five formulations.

---

[9] Meyer, C. A., McClintock, R. B., Silvestri, G. J., and Spencer, R. C., Jr., *Thermodynamic and Transport Properties of Steam*, American Society of Mechanical Engineers, 1967.

[10] Wagner, W., and Pruß, A., "The IAPWS Formulation 1995 for the Thermodynamic Properties of Ordinary Water Substance for General and Scientific Use," Journal of Physical Chemistry, Ref. Data 31, pp. 387-535, 2002.

[11] Research and Technology Committee on Water and Steam in Thermal Power Systems, *ASME Steam Properties for Industrial Use*, The American Society of Mechanical Engineers.

[12] Dalton, S., Excel Add-in Development in C/C++: Applications in Finance, John Wiley & Sons, Ltd., Chichester, England, 2005.

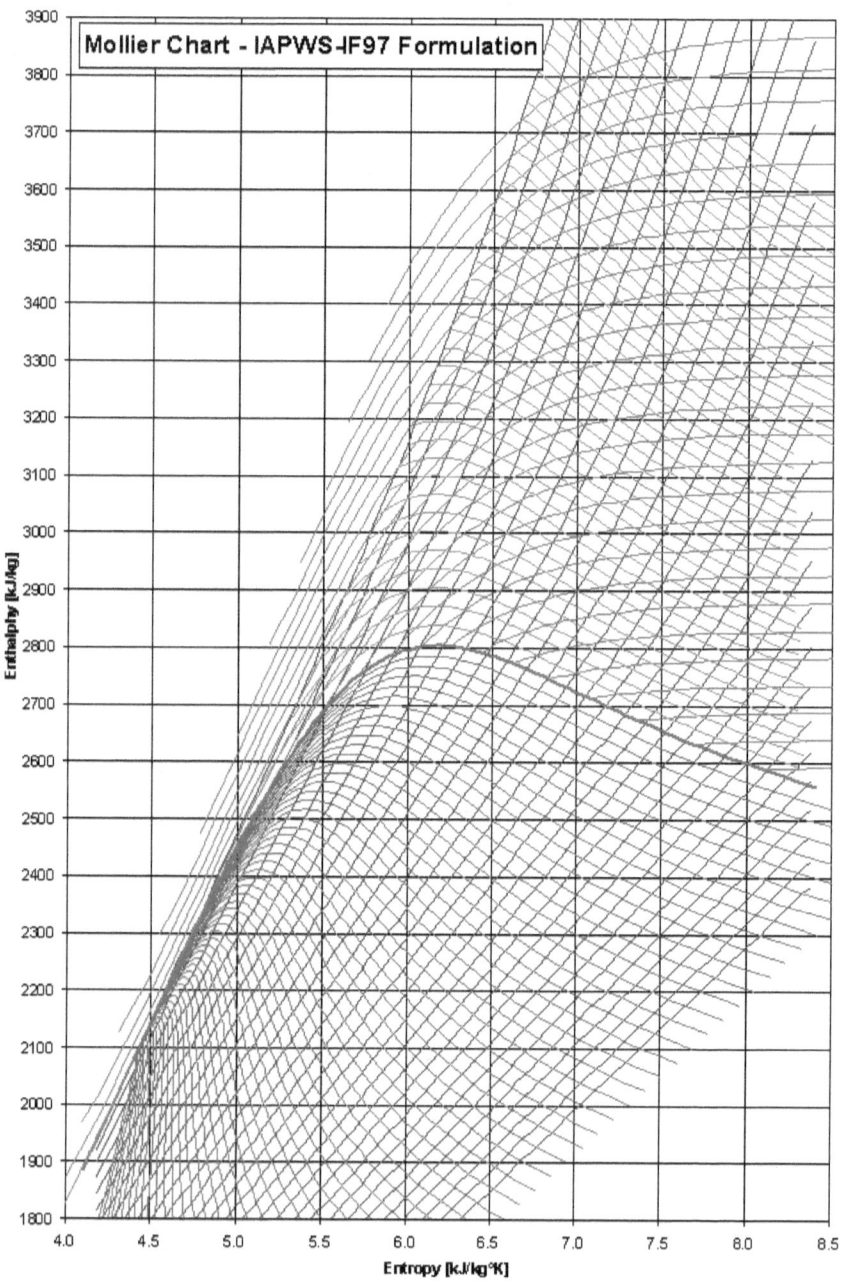

Mollier Chart - IAPWS-IF97 Formulation

Enthalphy [kJ/kg]

Entropy [kJ/kg°K]

90

Appendix B: Exhaust Properties

We will only consider complete combustion, that is, combustion with more than ample excess air, which is characteristic of gas turbine exhaust and duct burner operation. While there are small levels of NOx and SOx present in some HRSGs, these contributions have negligible impact on the thermal performance, which is the only aspect of HRSGs with which we are concerned. We are not, for instance considering corrosion (material) or mechanical (structural) factors.

Gas properties are defined in the NASA Glenn report.[13] Any other references (e.g., ASME PTC22) are derived from these. You will find spreadsheets and source code in folder examples\exhaust. Note that specific heat and enthalpy are mass-weighted, not mole-weighted, and GT exhaust composition is most often specified in mole fractions. The simplest calculations for specific heat and enthalpy are listed on the next page after the figure.

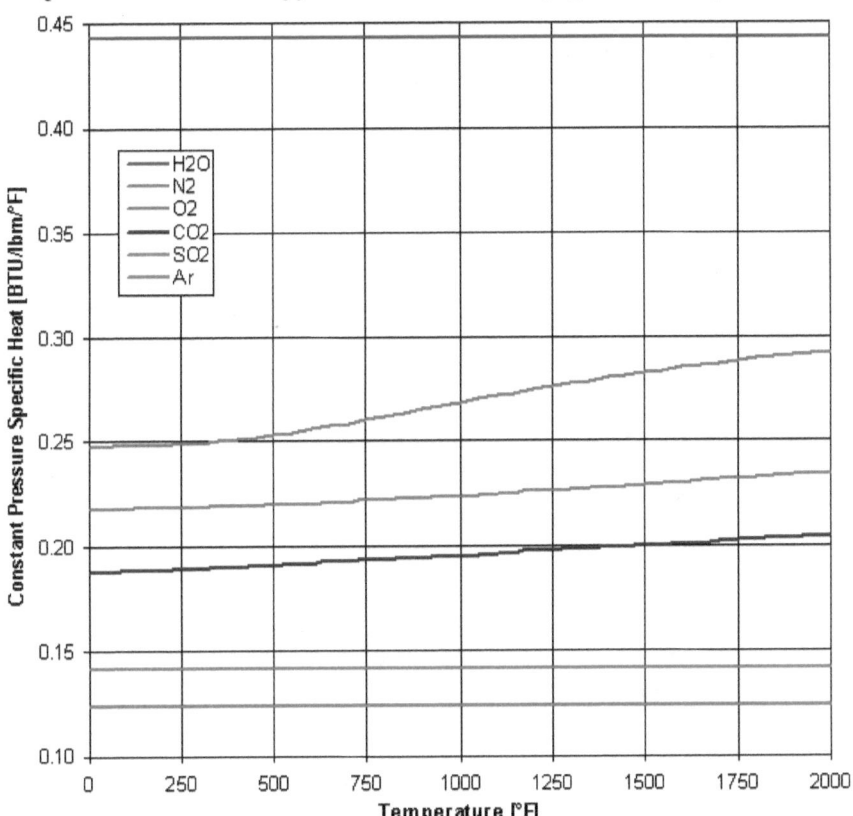

[13] McBride, B. J., Zehe, M. J., Gordon, S., "NASA Glenn Coefficients for Calculating Thermodynamic Properties of Individual Species," NASA Report No. 211556, 2002.

```
typedef struct{double a,b,c,d,MW;}GAS;
GAS N2 ={-1.823844E-01,2.450080E-01,
 9.515373E-06,1.278920E-09,28.013};
GAS O2 ={-1.173314E-02,2.177264E-01,
 1.570874E-06,1.001867E-09,31.998};
GAS CO2={ 1.374508E-03,1.873375E-01,
 3.953234E-06,1.303025E-10,44.010};
GAS H2O={ 9.393189E-05,4.431567E-01,
 6.169783E-08,9.216964E-12,18.015};
GAS Ar ={ 7.922561E-14,1.242788E-01,
-2.526427E-18,9.458965E-22,39.948};
GAS SO2={ 1.495539E-05,1.417236E-01,
 1.931810E-08,1.479716E-12,64.066};

double Hgas(MF mf,double T)
  {
  double HN2,HO2,HCO2,HH2O,HAr,HSO2,X,Y;
  HN2 =(( N2.d*T+ N2.c)*T+ N2.b)*T+ N2.a;
  HO2 =(( O2.d*T+ O2.c)*T+ O2.b)*T+ O2.a;
  HCO2=((CO2.d*T+CO2.c)*T+CO2.b)*T+CO2.a;
  HH2O=((H2O.d*T+H2O.c)*T+H2O.b)*T+H2O.a;
  HAr =(( Ar.d*T+ Ar.c)*T+ Ar.b)*T+ Ar.a;
  HSO2=((SO2.d*T+SO2.c)*T+SO2.b)*T+SO2.a;
  X= HN2* N2.MW*mf.N2
   + HO2* O2.MW*mf.O2
   +HCO2*CO2.MW*mf.CO2
   +HH2O*H2O.MW*mf.H2O
   + HAr* Ar.MW*mf.Ar
   +HSO2*SO2.MW*mf.SO2;
  Y= N2.MW*mf.N2
   + O2.MW*mf.O2
   +CO2.MW*mf.CO2
   +H2O.MW*mf.H2O
   + Ar.MW*mf.Ar
   +SO2.MW*mf.SO2;
  return(X/Y);
  }
double Tgas(MF mf,double H)
  {
  int iter;
  double T,T1,T2;
  T1=0.;
  T2=2000.;
  for(iter=0;iter<32;iter++)
    {
    T=(T1+T2)/2.;
    if(Hgas(mf,T)<H)
      T1=T;
    else
```

```
        T2=T;
    }
    return(T);
}
```

These calculations and additional spreadsheets can be found in the online archive in folder examples\exhaust. The most common gases are:

	°K	°C	°F	specific heat [kJ/kg/°C]					enthalpy [kJ/kg]				
				N2	O2	CO2	H2O	Ar	N2	O2	CO2	H2O	Ar
ambient	275	2	35	1.04	0.92	0.82	1.86	0.52	-24	-21	-8961	-13466	-12
	300	27	80	1.04	0.92	0.85	1.86	0.52	2	2	-8940	-13420	1
	325	52	125	1.04	0.92	0.87	1.87	0.52	28	25	-8918	-13373	14
HRSG gas	350	77	170	1.04	0.93	0.89	1.88	0.52	54	48	-8896	-13326	27
	375	102	215	1.04	0.93	0.92	1.89	0.52	80	71	-8874	-13279	40
	400	127	260	1.04	0.94	0.94	1.90	0.52	106	95	-8850	-13232	53
	425	152	305	1.05	0.95	0.96	1.91	0.52	132	118	-8827	-13184	66
	450	177	350	1.05	0.96	0.98	1.93	0.52	158	142	-8803	-13136	79
	475	202	395	1.05	0.96	1.00	1.94	0.52	185	166	-8778	-13088	92
	500	227	440	1.06	0.97	1.01	1.96	0.52	211	190	-8753	-13039	105
	525	252	485	1.06	0.98	1.03	1.97	0.52	237	215	-8727	-12990	118
	550	277	530	1.06	0.99	1.05	1.99	0.52	264	239	-8701	-12940	131
	575	302	575	1.07	1.00	1.06	2.00	0.52	291	264	-8675	-12891	144
	600	327	620	1.07	1.00	1.08	2.02	0.52	317	289	-8648	-12840	157
	625	352	665	1.08	1.01	1.09	2.03	0.52	344	314	-8621	-12790	170
	650	377	710	1.09	1.02	1.10	2.05	0.52	372	339	-8594	-12739	183
	675	402	755	1.09	1.02	1.11	2.06	0.52	399	365	-8566	-12687	196
	700	427	800	1.10	1.03	1.13	2.08	0.52	426	391	-8538	-12636	209
GT exhaust	725	452	845	1.10	1.04	1.14	2.10	0.52	454	416	-8510	-12583	222
	750	477	890	1.11	1.04	1.15	2.12	0.52	481	443	-8481	-12531	235
	775	502	935	1.12	1.05	1.16	2.13	0.52	509	469	-8452	-12478	248
	800	527	980	1.12	1.05	1.17	2.15	0.52	537	495	-8423	-12424	261
	825	552	1025	1.13	1.06	1.18	2.17	0.52	565	521	-8394	-12370	274
	850	577	1070	1.13	1.06	1.19	2.18	0.52	593	548	-8364	-12316	287
	875	602	1115	1.14	1.07	1.20	2.20	0.52	622	575	-8334	-12261	300
	900	627	1160	1.15	1.07	1.20	2.22	0.52	650	601	-8304	-12206	313

Table title: NASA Glenn Properties

The offset in the gas enthalpies is the heat of formation, which we don't want in our HRSG enthalpy but do want in the combustion calculations.

Appendix C: Combustion Calculations

We will only cover those aspects of combustion essential to HRSG modeling, which also means that we will only cover oil- and gas-fired duct burners. We will not cover gas turbine combustion, as the products of combustion are provided by the engine manufacturer and may account for inlet humidification (evaporative cooling or fogging), steam power augmentation, and water injection to control NOx formation.

Complete combustion of hydrocarbon fuels with moist air having excess oxygen is typical of gas turbines. The chemical reactions are often generalized in terms of the average fuel composition and the H/C molar ratio (m/n in this case). This reaction can be written:

$$0.7808N_2 + 0.2095O_2 + 0.0004CO_2 + 0.0093Ar + wH_2O$$
$$+ a(Cn + Hm) = \alpha N_2 + \beta O_2 + \gamma CO_2 + \delta Ar + \varepsilon H_2O \tag{C.1}$$

$$\alpha = 0.7808$$
$$\delta = 0.0093 \tag{C.2}$$

$$\beta = 0.2095 - a\left(n + \frac{m}{2}\right)$$
$$\gamma = 0.0004 + an \tag{C.3}$$
$$\varepsilon = w + \frac{m}{2}$$

The mole fractions of the constituents can be calculated directly from these parameters. The humidity ratio must be modified to account for the differing molecular weights of dry air and water vapor. The ambient and combustion functions are listed below:

```
typedef struct{double N2,O2,CO2,H2O,Ar;}MF;

MF Ambient(double baro,double Tdb,double RH)
  {
  double W,x,y;
  static MF mf;
  W=fWdbrh(baro,Tdb,RH);
  x=W/(1.+W);
  y=x*28.9645/18.01534;
  mf.H2O=y;
  mf.N2 =0.7808*(1.-y);
  mf.O2 =0.2095*(1.-y);
  mf.CO2=0.0004*(1.-y);
  mf.Ar =0.0093*(1.-y);
  return(mf);
  }
```

```
MF Combustion(MF mf1,double Fair,double Ffuel,double
    HCratio)
  {
  double C,H,S;
  static MF mf2;
  H=HCratio/(1.+HCratio);
  C=1.-H;
  mf2.N2 =Fair*mf1.N2;
  mf2.O2 =Fair*mf1.O2 -Ffuel*(C+H/2.);
  mf2.CO2=Fair*mf1.CO2+Ffuel*C;
  mf2.Ar =Fair*mf1.Ar;
  mf2.H2O=Fair*mf1.H2O+Ffuel*H/2.;
  S=mf2.N2+mf2.O2+mf2.CO2+mf2.Ar+mf2.H2O;
  mf2.N2 /=S;
  mf2.O2 /=S;
  mf2.CO2/=S;
  mf2.Ar /=S;
  mf2.H2O/=S;
  return(mf2);
  }
```

These calculations and additional spreadsheets can be found in the online archive in folder examples\exhaust and also examples\CCPP.

Appendix D. Moist Air Properties

The only properties of moist air considered are those developed by Hyland & Wexler[14,15,16], and refined by Nelson & Sauer.[17] The more recent formulation developed by Hermann, Kretzschmar, and Gatley[18,19] do not constitute a substantive improvement, merely an academic one; so there is little point implementing these. Moist air properties appear in various editions of the *ASHRAE Handbook of Fundamentals*. Beware that the equations in many editions of this otherwise excellent reference are wrong in that the equations contained therein don't produce the tabulated results.

In 1984 this author was part of a Cooling Technology Institute (CTI) task force investigating discrepancies in the published properties of moist air. The National Bureau of Standards (NBS)—now the National Institute of Standards and Technology (NIST)—lost Hyland & Wexler's original reports; however, a copy still existed in the Library of Congress (LoC). A colleague, Al Feltzin, went to the LoC and made a photocopy of the original reports. The tabulated values in the ASHRAE handbook are correct, but not all of the equations are, especially before 1993.

Formulations consistent with Hyland & Wexler, along with code plus an Excel® Add-In can be found in my book, *Evaporative Cooling*, and on my web site listed in the foreword. The code, calculations, and spreadsheets can be found in the online archive in folder examples\moistair. The humidity ratio, W, is the ratio of the mass of water to mass of dry air and is given by:

$$W = \left(\frac{MW_{H2O}}{MW_{AIR}} \right) \left(\frac{fP_{SAT}}{P_{BARO} - fP_{SAT}} \right) \qquad (D.1)$$

Where MW_{H2O} and MW_{AIR} are the molecular weights of water and air, respectively. P_{SAT} and P_{BARO} are the saturation and barometric pressures,

---

[14] Hyland, R. W., Wexler, A., and Stewart, R., "Thermodynamic Properties of Dry Air, Moist Air and Water and SI Psychrometric Charts," ASHRAE RP-216 and RP-25, 1983.

[15] Hyland, R. W. and Wexler, A., "Formulations for the Thermodynamic Properties of the Saturated Phases of H2O from 173.15 K to 473.15 K," ASHRAE Trans., Vol. 89, pp. 500-519, 1983.

[16] Hyland, R. W. and Wexler, A., "Formulations for the Thermodynamic Properties of Dry Air from 173.15 K to 473.15 K, and of Saturated Moist Air from 173.15 K to 372.15 K, at Pressures to 5 MPa," ASHRAE Trans., Vol. 89, pp. 520-535, 1983.

[17] Nelson, H. F. and Sauer, H. J., "Formulation of High-Temperature Properties for Moist Air," HVAC&R Research Vol. 8, pp. 311-334, 2002.

[18] Herrmann, S., Kretzschmar, H.-J., and Gatley, D. P., "Thermodynamic Properties of Real Moist Air, Dry Air, Steam, Water, and Ice," HVAC&R Research, 2009.

[19] Herrmann, S., Kretzschmar, H.-J., and Gatley, D. P., "Thermodynamic Properties of Real Moist Air, Dry Air, Steam, Water, and Ice - Final Report," ASHRAE RP-1485, 2009.

respectively. The enhancement factor, f, is the ratio of the effective partial pressure of water vapor in air at the saturation point to the saturation pressure of steam alone (no air present). The enhancement factor varies with temperature and barometric pressure and is shown in the figure below for one atmosphere:

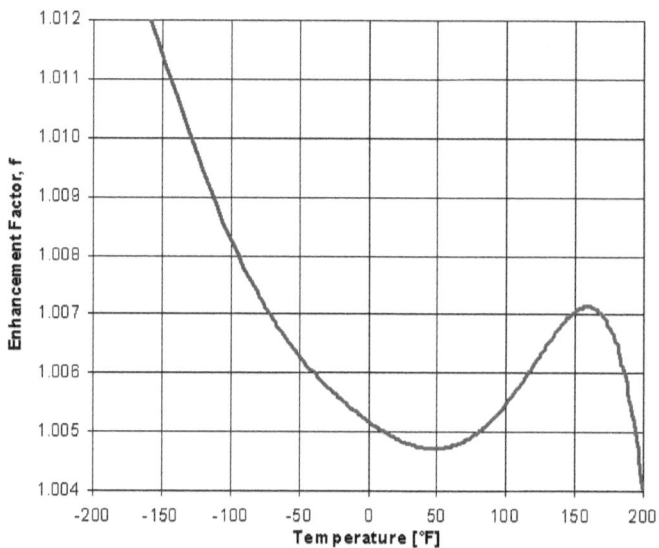

The saturation pressure for water vapor (against water liquid without air present) is shown in this next figure:

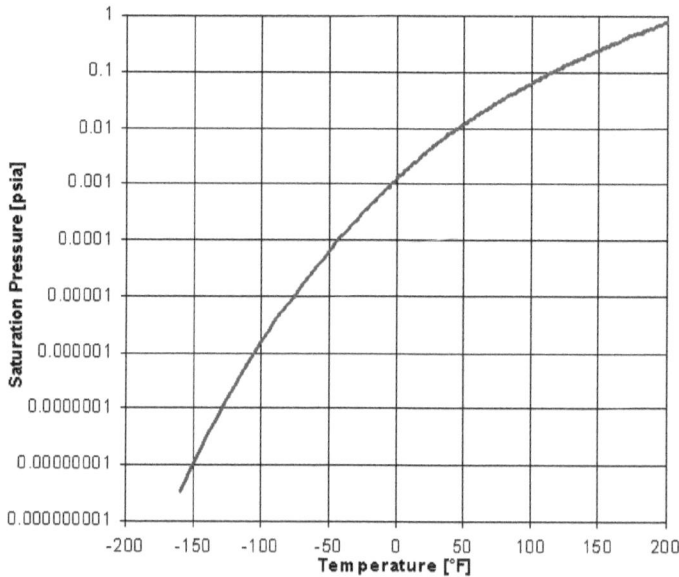

The resulting humidity ratio is then:

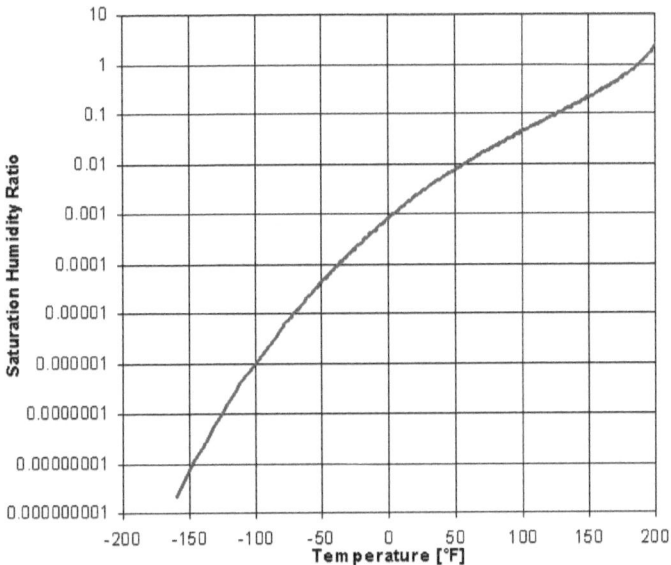

The enthalpy of moist air (per pound of dry air) is given by:

$$h = h_A + Wh_G = 0.24\,T + W(1061 + 0.444\,T) \qquad (D.2)$$

where h_A and h_G are the enthalpies of dry air and water vapor, respectively. The enthalpy of water vapor is given by:

$$h_G = 1061 + 0.444\,T \qquad (D.3)$$

Note that this is the enthalpy of saturated water vapor, not the latent heat (h_{FG}) as is sometimes thought and even cited in the literature.[20]

Note that you will need rigorous psychrometric calculations at the gas turbine (GT) inlet, but DO NOT use these in the HRSG, as they are meaningless above the boiling point (212°F/100°C). In that case, use NASA Glenn, but be aware that the references are different. ASHRAE (Hyland & Wexler) reference liquid water at the triple point and NASA Glenn reference vapor water at standard conditions. If you do not consider this difference in the energy balance around a GT, you will get erroneous results.

---

[20] Need proof? Consider this... if you were to add steam at the critical point to dry air, you would significantly increase the enthalpy of the mixture; however, at the critical point h_{FG} is zero.

Appendix E. Heat Transfer Coefficients

Aside from tube wall resistance, which is trivial to calculate, the most important contributors to the overall heat transfer coefficient, U, are the inside and outside convective transports plus boiling in the case of evaporators. Fouling resistance is generally assumed, based on typical observed values, which are readily available. These conductances are combined as reciprocal resistances using Equation 3.1.

Convective Heat Transfer Coefficients

The convective contributions inside and outside the tubes are calculated similarly, only different correlations and properties are used for the gas and steam sides. These correlations all take the form of the classic equation, as developed by Dittus&Boelter[21]:

$$Nu = \frac{hD}{k} = 0.023 \, \mathrm{Re}^{\frac{4}{5}} \, \mathrm{Pr}^{n} \tag{E.1}$$

where n=0.4 for heating and n=0.3 for cooling. A more recent correlation has been developed by Sieder&Tate[22]:

$$Nu = 0.027 \, \mathrm{Re}^{\frac{4}{5}} \, \mathrm{Pr}^{\frac{1}{3}} \left(\frac{\mu_{BULK}}{\mu_{SURFACE}} \right)^{0.14} \tag{E.2}$$

Yet another correlation has been developed by Rabas&Cane[23]:

$$Nu = 0.0158 \, \mathrm{Re}^{0.835} \, \mathrm{Pr}^{0.462} \tag{E.3}$$

Resistance of the tube wall can be found in any heat transfer text:

$$R_W = \frac{D_O}{2k_W} \ln\left(\frac{D_O}{D_I} \right) \tag{E.4}$$

There are numerous correlations for the shell side heat transfer coefficients. There is considerable discussion in the literature on which correlation should be preferred and I suspect that each manufacturer has their own equation that has been fine-tunde. These vary with the fluids, baffles, and tube spacing. Fouling is often used as a *fudge* factor. The overall heat transfer coefficient is found by summing the resistances:

---

[21] Dittus, P. W. and L. M. Boelter, L.M., University of California Publications in Engineering, Vol. 1, No. 13, pp. 443-461 1930 (reprinted in *International Communications in Heat and Mass Transfer*, Vol. 12, pp. 3-22, 1985).

[22] Sieder, E. N. and G. E. Tate, "Heat Transfer and Pressure Drop of Liquids in Tubes," Industrial Engineering Chemistry, Vol. 28, p. 1429, 1936.

[23] Rabas, T. J., and D. Cane, "An Update of Intube Forced Convection Heat Transfer Coefficients of Water," *Desalinization*, Vol. 44, pp. 109–119, 1983.

$$\frac{1}{U} = \left(\frac{D_O}{D_I}\right)(R_{FI} + R_T) + R_M + R_S + R_{FO} \qquad (E.5)$$

Here R_{FI} and R_{FO} are the inside and outside fouling resistance, respectively. R_T, R_W, and R_S are the tube side, tube material, and shell side resistances, respectively. The D_O/D_I term accounts for the fact that the inner and outer surface areas of a tube aren't the same per unit length.

There are also several methods for estimating the shell side heat transfer coefficient. One of the earliest methods was developed by Kern[24], based on industrial heat exchangers and is quite similar to the Sieder-Tate.

$$Nu = 0.36 \, Re^{0.55} \, Pr^{\frac{1}{3}} \left(\frac{\mu_{BULK}}{\mu_{SURFACE}}\right)^{0.14} \qquad (E.6)$$

This correlation could also be cast in the same form as Dittus-Boelter for convenience:

$$Nu = 0.36 \, Re^{0.55} \, Pr^{n} \qquad (E.7)$$

Tube Bundles

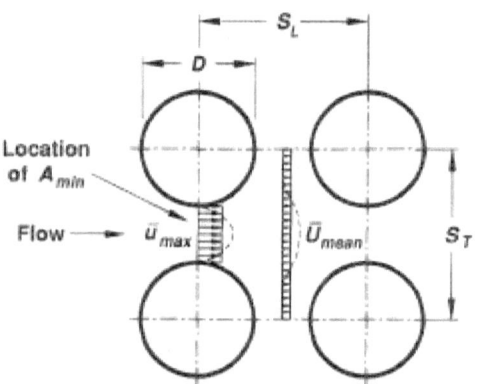

Inline Tube Bank

The length in both the Nusselt and Reynolds number is an equivalent diameter (D_E), that takes into account the tube spacing, pitch (P), and bundle alignment. This first equation (E.7) is for a square tube arrangement:

$$D_E = \frac{4P^2}{\pi D_O} - D_O \qquad (E.8)$$

---

[24] Kern, D. Q., *Process Heat Transfer*, McGraw-Hill, 1950.

and this second (E.8) is for a triangular arrangement:

$$D_E = \frac{2\sqrt{3}P^2}{\pi D_O} - D_O \qquad \text{(E.9)}$$

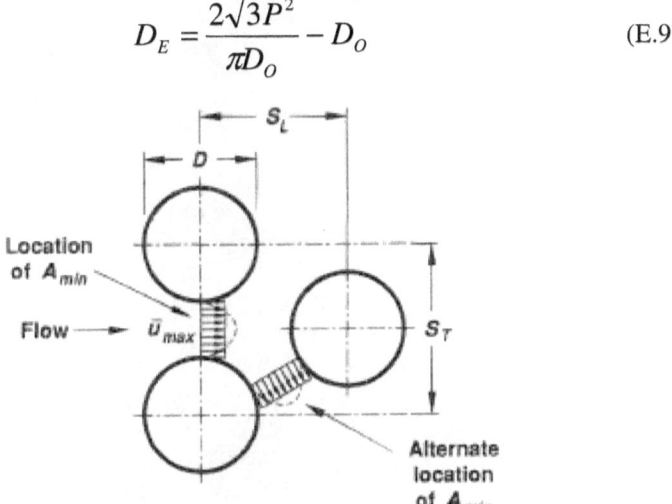

Staggered Tube Bundle

The shell-side velocity, V_s, is given by:

$$V_S = \frac{\dot{m}_S}{\rho N_R (P - D_O)\left(\dfrac{L_T}{N_B}\right)} \qquad \text{(E.10)}$$

where N_R is the number of tubes per row (across the flow), L_T is the tube length, and N_B is the number of baffles. You may want to adjust the effective shell-side area to better represent the actual tube bundle and baffle arrangement in your heat exchanger. The Reynolds and Prandtl numbers are given by:

$$\text{Re} = \frac{DV}{\upsilon} \qquad \text{(E.11)}$$

$$\text{Pr} = \frac{\mu C}{k} \qquad \text{(E.12)}$$

You can devote considerable effort to calculating and refining these heat transfer coefficients, but always remember that the end result should be in the range discussed in Chapter 3. Years of performance testing experience have proven that there are limits to what may be practically achieved.

Various Tube Bundles

<u>Finned Tubes</u>

An excellent source of information for finned tubes is the Wolverine Engineering Data Book III.[25] This reference contains a variety of correlations, which you may find useful. Many correlations for convection over finned tubes are inspired by the Reynolds Analogy, which states that the Stanton Number is equal to the friction factor divided by two:

$$St = \frac{f}{2} \tag{E.13}$$

The Stanton number is also equal to the Nusselt number divided by the Reynolds and Prandtl numbers.

$$St = \frac{Nu}{\text{Re}\,\text{Pr}} = \frac{h}{\rho u C_P} \tag{E.14}$$

Fins do increase the heat transfer coefficient and also the pressure drop. Fin designs vary considerably. One common design is shown on the next page.

---

[25] Thome, J. R. *Engineering Data Book III*, Wolverine Tube, Inc. 2009.

Finned Tubes

Appendix F. Flow Measurement

The only accurate flow measurement for HRSG testing is either a Coriolis meter or sharp-faced orifice for fuel gas flow or an ASME-grade calibrated nozzle for feedwater. Performing a test based on an uncalibrated and/or unverifiably clean nozzle is a waste of time and money. Do not bother measuring steam flow in the vapor phase. The applicable standards for such calculations are either ASME MFC-3M (1989 or 2004), "Measurement of Fluid Flow in Pipes Using Orifice, Nozzle, and Venturi" and ASME PTC-19.5 (2004), "Flow Measurement." These standards contain various corrections and correlations for each of the pertinent devices. There are also ranges of applicability for physical dimensions, for instance, diameter ratios. Do not exceed these. The equations will not be duplicated here, as there are numerous considerations and figures that must be considered and these details are beyond the scope of this text.

Make sure that the range of operating conditions are within the applicability for each flow device. More than once I have encountered an otherwise well-designed system that could not be tested within the range of calibration of the feedwater flow nozzle and still demonstrate the guarantees. I have also encountered nozzles that were calibrated at Reynolds numbers far removed from the intended installation. Yet other cases exist where a very expensive nozzle ($15K) was never calibrated. Cleanliness is also important. It is often presumed that feedwater nozzles are clean, but this may not be true and these should be inspected. Physical damage (e.g., scraping, gouging, dents, etc.) can be sustained during shipping, storage, and installation so that scaling and corrosion are not the only things to consider. More than once I have seen a sharp-faced orifice installed backwards. The sharp face goes into the flow, as shown below:

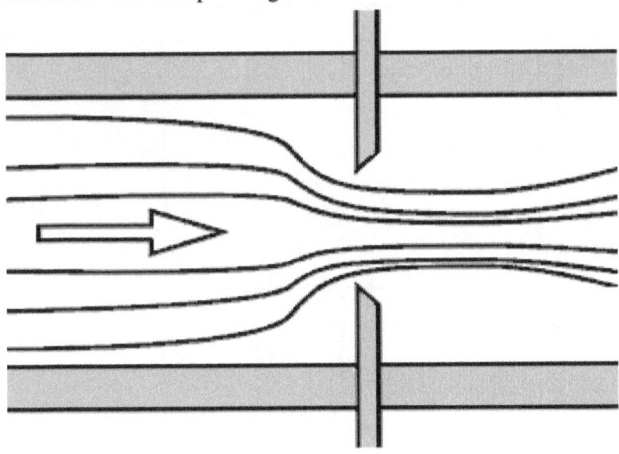

The calculation of flow can be quite involved. The following is typical for fuel gas through an orifice.

Input Data	Units	Test1	Test2	Test3	Avg.
Meter Type		ORIFICE	ORIFICE	ORIFICE	ORIFICE
Pipe Material		SS3	SS3	SS3	SS3
Throat Material		SS3	SS3	SS3	SS3
Pipe Diameter	inches	7.9810	7.9810	7.9810	7.9810
Throat Diameter	inches	4.7886	4.7886	4.7886	4.7886
Beta Ratio		0.6000	0.6000	0.6000	0.6000
Flowing Pressure	psia	483.40	483.32	483.17	483.30
Flowing Temperature	Deg F	25.69	25.41	26.21	25.77
Differential Pressure	in. H2O	120.63	120.13	120.24	120.34
Frequency	pulse	120.63	120.13	120.24	120.34
Fluid Type		GAS	GAS	GAS	GAS
Pipe Alpha Coefficient		9.000E-06	8.999E-06	9.001E-06	9.000E-06
Throat Alpha Coefficient		9.000E-06	8.999E-06	9.001E-06	9.000E-06
Pipe Reynolds Number		5,431,835	5,420,046	5,406,809	5,419,563
Corrected Pipe Diameter	inches	7.9775	7.9775	7.9776	7.9775
Corrected Throat Diameter	inches	4.7865	4.7865	4.7865	4.7865
Corrected Beta Ratio		0.6000	0.6000	0.6000	0.6000
Tap Term (for use in AGA 3 extrap. Cd)		-0.000209	-0.000209	-0.000209	-0.0002
A Term (for use in AGA 3 extrap. Cd)		0.007204	0.007216	0.007231	0.0072
C Term (for use in AGA 3 extrap. Cd)		0.553055	0.553476	0.553950	0.5535
Discharge Coefficient (ASME MFC 3M)		0.6045	0.6045	0.6045	0.6045
Discharge Coefficient (AGA3)		0.6041	0.6041	0.6041	0.6041
Discharge Coefficient (Cal. AGA R3)		0.6031	0.6031	0.6031	0.6031
Discharge Coefficient (Cal. PTC 19.5)		0.6041	0.6041	0.6041	0.6041
Discharge Coefficient		0.6045	0.6045	0.6045	0.604
Critical Pressure, Pcr	psia	667.95	668.29	666.98	667.74
Critical Temperature, Tcr	Deg F	346.64	346.49	345.89	346.34
Critical Volume, Vcr	ft³/lb	0.0980	0.0980	0.0981	0.0980
Specific Gravity Rel. To Air, Gideal	ratio	0.5726	0.5717	0.5719	0.5721
Base Pressure, Pbase	psia	14.69	14.69	14.69	14.69
Base Temperature, Tbase	Deg F	70.00	70.00	70.00	70.00
AGA 8 Calculated Fuel Density	lb / ACF	1.6826	1.6802	1.6760	1.6796
AGA 8 Calculated Base Fuel Density	lb / SCF	0.0439	0.0438	0.0438	0.0439
Gas Viscosity	lbm/ft-sec	7.411E-06	7.407E-06	7.419E-06	7.412E-06
Natural Gas Flow Rate using NX-19 Density	KSCFH	1,761.83	1,760.00	1,757.99	1,759.94
Natural Gas Flow Rate using AGA8 Density	KSCFH	1,723.94	1,722.00	1,720.12	1,722.02
Natural Gas Flow Rate using AGA8 Density	KPPH	75.67	75.46	75.40	75.51

Nozzle flow calculations can also be quite involved, as illustrated in the following table:

Description	units	Design	Test1	Test2	Test3	Avg/
INPUTS						
Meter Type		NOZZLE	NOZZLE	NOZZLE	NOZZLE	NOZZLE
Pipe Material		CS	CS	CS	CS	CS
Throat Material		SS3	SS3	SS2	SS3	SS4
Tap Set		1.00	1.00	1.00	1.00	1.00
Pipe Diameter	inches	18.3771	18.3771	18.3771	18.3771	18.3771
Throat Diameter	inches	8.8875	8.8875	8.8875	8.8875	8.8875
Flowing Temperature	Deg F	270.0	270.0	270.0	276.0	272.0
Flowing Pressure	psia	350.0	350.0	350.0	348.0	349.3
Differential Pressure	in. H2O	377.7	398.8	378.1	372.0	383.0
Frequency	pulse	0.00	0.00	-1.00	0.00	-0.33
Fluid Type		WATER	WATER	WATER	WATER	WATER
Viscosity	lbm/ft-sec	1.631E-06	1.631E-06	1.631E-06	1.588E-06	1.617E-06
Fluid Density (1997)	lb/ft³	58.3106	58.3106	58.3106	58.1301	58.2504
Reynolds Number, Re	-	6.16E+08	6.33E+08	6.15E+08	6.27E+08	6.25E+08
Gas Expansion Factor, Y1	-	1.0000	1.0000	1.0000	1.0000	1.0000
Beta @ Flowing Conditions	-	0.4839	0.4839	0.4836	0.4839	0.4838
Disc. Coeff. - PTC 19.5	-	1.0025	1.0025	1.0025	1.0025	1.0025
Disc. Coeff. (manual calc.)	-	0.9973	0.9973	0.9973	0.9973	0.9973
RESULTS						
Flow Rate (method 1)	lb/hr	4,354,496	4,474,396	4,351,473	4,315,146	4,380,338
Flow Rate (method 2)	lb/hr	4,355,040	4,474,956	4,352,017	4,315,680	4,380,884

also by D. James Benton

3D Articulation: Using OpenGL, ISBN-9798596362480, Amazon, 2021 (book 3 in the 3D series).

3D Models in Motion Using OpenGL, ISBN-9798652987701, Amazon, 2020 (book 2 in the 3D series.

3D Rendering in Windows: How to display three-dimensional objects in Windows with and without OpenGL, ISBN-9781520339610, Amazon, 2016 (book 1 in the 3D series).

A Synergy of Short Stories: The whole may be greater than the sum of the parts, ISBN-9781520340319, Amazon, 2016.

Azeotropes: Behavior and Application, ISBN-9798609748997, Amazon, 2020.

bat-Elohim: Book 3 in the Little Star Trilogy, ISBN-9781686148682, Amazon, 2019.

Boilers: Performance and Testing, ISBN: 9798789062517, Amazon 2021.

Combined 3D Rendering Series: 3D Rendering in Windows®, 3D Models in Motion, and 3D Articulation, ISBN-9798484417032, Amazon, 2021.

Complex Variables: Practical Applications, ISBN-9781794250437, Amazon, 2019.

Compression & Encryption: Algorithms & Software, ISBN-9781081008826, Amazon, 2019.

Computational Fluid Dynamics: an Overview of Methods, ISBN-9781672393775, Amazon, 2019.

Computer Simulation of Power Systems: Programming Strategies and Practical Examples, ISBN-9781696218184, Amazon, 2019.

Contaminant Transport: A Numerical Approach, ISBN-9798461733216, Amazon, 2021.

CPUnleashed! Tapping Processor Speed, ISBN-9798421420361, Amazon, 2022.

Curve-Fitting: The Science and Art of Approximation, ISBN-9781520339542, Amazon, 2016.

Death by Tie: It was the best of ties. It was the worst of ties. It's what got him killed., ISBN-9798398745931, Amazon, 2023.

Differential Equations: Numerical Methods for Solving, ISBN-9781983004162, Amazon, 2018.

Equations of State: A Graphical Comparison, ISBN-9798843139520, Amazon, 2022.

Evaporative Cooling: The Science of Beating the Heat, ISBN-9781520913346, Amazon, 2017.

Forecasting: Extrapolation and Projection, ISBN-9798394019494, Amazon 2023.

Heat Engines: Thermodynamics, Cycles, & Performance Curves, ISBN-9798486886836, Amazon, 2021.

Heat Exchangers: Performance Prediction & Evaluation, ISBN-9781973589327, Amazon, 2017.

Heat Transfer: Heat Exchangers, Heat Recovery Steam Generators, & Cooling Towers, ISBN-9798487417831, Amazon, 2021.

Heat Transfer Examples: Practical Problems Solved, ISBN-9798390610763, Amazon, 2023.

The Kick-Start Murders: Visualize revenge, ISBN-9798759083375, Amazon, 2021.

Jamie2: Innocence is easily lost and cannot be restored, ISBN-9781520339375, Amazon, 2016-18.

Kyle Cooper Mysteries: Kick Start, Monte Carlo, and Waterfront Murders, ISBN-9798829365943, Amazon, 2022.

The Last Seraph: Sequel to Little Star, ISBN-9781726802253, Amazon, 2018.

Little Star: God doesn't do things the way we expect Him to. He's better than that! ISBN-9781520338903, Amazon, 2015-17.

Living Math: Seeing mathematics in every day life (and appreciating it more too), ISBN-9781520336992, Amazon, 2016.

Lost Cause: If only history could be changed..., ISBN-9781521173770, Amazon, 2017.

Mass Transfer: Diffusion & Convection, ISBN-9798702403106, Amazon, 2021.

Mill Town Destiny: The Hand of Providence brought them together to rescue the mill, the town, and each other, ISBN-9781520864679, Amazon, 2017.

Monte Carlo Murders: Who Killed Who and Why, ISBN-9798829341848, Amazon, 2022.

Monte Carlo Simulation: The Art of Random Process Characterization, ISBN-9781980577874, Amazon, 2018.

Nonlinear Equations: Numerical Methods for Solving, ISBN-9781717767318, Amazon, 2018.

Numerical Calculus: Differentiation and Integration, ISBN-9781980680901, Amazon, 2018.

Numerical Methods: Nonlinear Equations, Numerical Calculus, & Differential Equations, ISBN-9798486246845, Amazon, 2021.

Orthogonal Functions: The Many Uses of, ISBN-9781719876162, Amazon, 2018.

Overwhelming Evidence: A Pilgrimage, ISBN-9798515642211, Amazon, 2021.

Particle Tracking: Computational Strategies and Diverse Examples, ISBN-9781692512651, Amazon, 2019.

Plumes: Delineation & Transport, ISBN-9781702292771, Amazon, 2019.

Power Plant Performance Curves: for Testing and Dispatch, ISBN-9798640192698, Amazon, 2020.

Practical Linear Algebra: Principles & Software, ISBN-9798860910584, Amazon, 2023.

Props, Fans, & Pumps: Design & Performance, ISBN-9798645391195, Amazon, 2020.

Remediation: Contaminant Transport, Particle Tracking, & Plumes, ISBN-9798485651190, Amazon, 2021.

ROFL: Rolling on the Floor Laughing, ISBN-9781973300007, Amazon, 2017.

Seminole Rain: You don't choose destiny. It chooses you, ISBN-9798668502196, Amazon, 2020.

Septillionth: 1 in 10^{24}, ISBN-9798410762472, Amazon, 2022.

Software Development: Targeted Applications, ISBN-9798850653989, Amazon, 2023.

Software Recipes: Proven Tools, ISBN-9798815229556, Amazon, 2022.

Steam 2020: to 150 GPa and 6000 K, ISBN-9798634643830, Amazon, 2020.

Thermochemical Reactions: Numerical Solutions, ISBN-9781073417872, Amazon, 2019.

Thermodynamic and Transport Properties of Fluids, ISBN-9781092120845, Amazon, 2019.

Thermodynamic Cycles: Effective Modeling Strategies for Software Development, ISBN-9781070934372, Amazon, 2019.

Thermodynamics - Theory & Practice: The science of energy and power, ISBN-9781520339795, Amazon, 2016.

Version-Independent Programming: Code Development Guidelines for the Windows® Operating System, ISBN-9781520339146, Amazon, 2016.

The Waterfront Murders: As you sow, so shall you reap, ISBN-9798611314500, Amazon, 2020.

Weather Data: Where To Get It and How To Process It, ISBN-9798868037894, Amazon, 2023.